The West Point Military Library

General Editors of the Series

COLONEL THOMAS E. GRIESS
Professor, United States Military Academy

and

PROFESSOR JAY LUVAAS
Allegheny College

GREENWOOD PRESS, in cooperation with the United States Military Academy and its Library, is proud to publish this extensive series of books and journals in the fields of military history and theory. The WEST POINT MILITARY LIBRARY series consists largely of clothbound, facsimile reprints of rare, important, and out-of-print books and journals, but it will also include a variety of original scholarly and professional publications. The series will focus on books and journals of outstanding interest and significance in military history, although works which have shaped American military thinking will be given special consideration. While most of the initial reprint and original works will be in English, the series also will include both foreign-language titles and original translations. Edited archival and documentary material (letters, diaries, field orders, commanders' notebooks, etc.) will form a separate part of the series.

Titles in The West Point Military Library

BARNARD, *Military Schools and Courses of Instruction in the Science and Art of War* (1872).
BIGELOW, *The Principles of Strategy. Illustrated Mainly from American Campaigns*, 2nd ed. (1894).
BIRKHIMER, *Historical Sketch of the Organization, Administration, Matériel and Tactics of the Artillery, United States Army* (1884).
BRACKETT, *History of the United States Cavalry to 1863* (1865).
FEUQUIERES, *Memoirs Historical and Military: Containing a Distinct View of All the Considerable States of Europe* (1735-1736). 2 vols.
HALLECK, *Elements of Military Art and Science* (1846).
HAMLEY, *The Story of the Campaign of Sebastopol, Written in the Camp* (1855).
HOYT, *Practical Instructions for Military Officers* (1811).
JAMES, *A New and Enlarged Military Dictionary, in French and English* (1810). 2 vols.
JOMINI, *Summary of the Art of War* (1854).
LLOYD, *The History of the Late War in Germany, Between the King of Prussia, and the Empress of Germany and Her Allies* (1781). 3 vols.
MAC DONALD, *Rules and Regulations for the Field Exercise and Manoeuvres of the French Infantry*, 2nd ed. (1806). 2 vols.
MAHAN, *The Story of the War in South Africa, 1899-1900 (1900).*
MAHAN, *A Complete Treatise on Field Fortification* (1836).
MALLESON, *The Battle-fields of Germany, from the Outbreak of the Thirty-Years War to the Battle of Blenheim* (1884).
MARMONT, *The Spirit of Military Institutions* (1864).
MITCHELL, *The Life of Wallenstein, Duke of Friedland* (1837).
MURRAY (ed.), *The Letters and Dispatches of John Churchill, First Duke of Marlborough, from 1702 to 1712* (1845). 5 vols.
The Officer's Manual in the Field; or, a Series of Military Plans, Representing the Principal Operations of a Campaign, 2nd ed. (1800).
PRUSSIA, Kriegsministerium, *Regulations for the Prussian Cavalry* (1757).
PRUSSIA, Kriegsministerium, *Regulations for the Prussian Infantry* (1759).
SCHELIHA, *A Treatise on Coast-Defense* (1868).
SCHMIDT, *Instructions for the Training, Employment, and Leading of Cavalry* (1881).
SCOTT, *Military Dictionary* (1862).
TOUSARD, *American Artillerist's Companion, or Elements of Artillery* (1809). 3 vols.
TURNER, *Pallas Armata. Military Essays of the Ancient Grecian, Roman, and Modern Art of War* (1683).
U.S. Military Academy, West Point, *The Centennial of the United States Military Academy at West Point, New York, 1802-1902* (1904). 2 vols.
U.S. War Department, *Instruction for Field Artillery* (1861).
UPTON, *The Armies of Asia and Europe* (1878).
UPTON, *Infantry Tactics, Double and Single Rank Adapted to American Topography and Improved Fire-arms*, rev. ed. (1874).
UPTON, *The Military Policy of the United States* (1904).
VERDY DU VERNOIS, *A Tactical Study, Based on the Battle of Custozza, 24th of June, 1866* (1894).
VERDY DU VERNOIS, *With the Royal Headquarters in 1870-71* (1897).
VIELE, *Handbook for Active Service* (1861).
WAGNER, *The Campaign of Königgrätz, a Study of the Austro-Prussian Conflict in the Light of the American Civil War* (1889).

A

COMPLETE TREATISE

ON

FIELD FORTIFICATION,

WITH

THE GENERAL OUTLINES

OF

THE PRINCIPLES

Regulating the Arrangement, the Attack, and the Defence

OF

Permanent Works.

BY D. H. MAHAN,
Professor of Military and Civil Engineering in the United States' Military Academy.

GREENWOOD PRESS, PUBLISHERS
NEW YORK

Originally published in 1836 by Wiley and Long

First Greenwood Reprinting, 1968

Library of Congress Catalogue Card Number: 68-54797

ISBN 978-0-8371-0557-4

PRINTED IN THE UNITED STATES OF AMERICA

TO

THE OFFICERS OF MILITIA

OF

THE UNITED STATES.

GENTLEMEN:—

THAT an efficiently organized Militia is the firmest, and only safe, bulwark of the State, is a political axiom admitted by all who understand the nature of our free institutions. Whatever, then, may contribute, in any degree, to strengthen this arm of our national defence, it is to be hoped will meet with an indulgent reception from the public. Feeling a strong conviction, that its efficiency must depend upon sound military instruction, I have presumed to dedicate the following work to you, gentlemen, in the hope that it may be found useful to our country, by proving serviceable to yourselves. Should this, my chief aim, be accomplished, I shall receive the highest gratification, in being thus enabled to make some slight return to my country, for the great benefit which she has conferred on me, in an education at our only National School.

With sentiments of high respect,

I have the honor to be,

GENTLEMEN,

Your obedient servant,

D. H. MAHAN.

UNITED STATES' MILITARY ACADEMY,
WEST POINT, JULY 14th, 1836.

PREFACE.

In preparing the work now laid before the public, which is chiefly designed for the use of the Cadets of the United States' Military Academy, the aim of the writer was to make a book which should also be generally useful,—one that should contain all the principles and important details of that branch of the *Art of Fortification* of which it specially treats, developed in a manner to be within the comprehension of any person of ordinary intelligence,—a book not for the study alone, but one which the officer can take with him into the camp, and consult at any moment.

With respect to the contents, the writer has collected his information from every source within his reach; rejecting all unnecessary discussions, and admitting no results, which have not either been submitted to a rational analysis, or tested by fair experiment; preferring, in all cases, the views of persons who wrote from their own experience, to the conjectures of others whose theories, however plausible, rested on a less solid foundation.

In arranging the matter, the writer has endeavored, as far as it was practicable, to confine the text to the principles alone, reserving the minor details for the plates, and the explanations appended to them; and on this point he would hazard one or two remarks.

Much that appertains to the Engineer's Art, is but an affair of feet and inches; facts which are the results of long usage, holding, in many instances, the important position of principles. His experience has taught him that those authors are the clearest who enter into the minutiæ of their subject; and that with pupils of superior minds, a thorough knowledge of details is an invaluable aid in unravelling the difficulties, and retaining the principles of the Art; whilst, with those of limited capacities, a want of such detail leaves them with the most vague and unsatisfactory notions.

Having said thus much of the plan of the work, a few words, respecting its object, may not be misplaced. The importance of a knowledge of Field Fortification, to officers of all arms in the regular army, need not be insisted upon; the daily occurrences of a campaign, demonstrate this too clearly. To the militia officer, this knowledge is even of more importance than to the regular; for called upon, in many cases, to act without the co-operation of regular troops, in the defence of his own fire-side, he will require all those conservative means which add strength and confidence to irregular forces when brought for the first time before an enemy.

To suppose irregular forces capable of coping on equal terms with disciplined troops, is to reason, not only against all probability, but against a vast weight of testimony to the contrary. It is not indeed that discipline confers individual courage; certainly a greater proportion of this essential military virtue will, in all cases, be found among the militia, composed as it must be of men of a higher grade of moral and intellectual qualities, than is to be met with among the common soldiery of any country. But these men, with all the superior advantages that must actuate the volunteer, necessarily want that indispensable element of success in pitched battles which results from discipline and habitual training. Called out on a particular emergency, with little or no previous exercise in the services they are required to render, militia have not that shoulder-to-shoulder courage, by which men are animated, who have served long together, which begets a reliance on each other, and a confidence in their chiefs, and which is one of the surest guaranties of victory. But place the militia soldier on his natural field of battle, behind a breast work, and an equilibrium between him and his more disciplined enemy is immediately established; with a feeling of security in his position, his confidence in his own exertions is restored; with a full certainty that his enemy cannot close upon him, before he can retire beyond his reach, he does his duty coolly, and with an execution so terrible, as to have placed the achievements of our militia, from the day of Bunker Hill, to the closing scene of our last war at New-Orleans, on a line with the most brilliant military exploits of the best disciplined troops in the world.

But were not these reasons in themselves sufficient, others of greater cogency could be adduced, in favor of intrenched positions for militia. Its ranks are filled with all that is most valuable in society. The farmer, the mechanic, the merchant, the members of the learned professions, must all quit their peaceful avocations to meet the foe. The father of the family, jeopards its future prosperity; the son, exposes his widowed mother to the chances of an old age of penury, to bare their breasts to a mercenary band, without other home, without other ties, than the camp affords. Surely nothing but a reckless disregard for the best interests of society, could urge men, under such circumstances, to forego the advantages of every possible conservative means.

The chapters of the work on Military Bridges, Reconnaissances, and Permanent Fortification, are necessarily incomplete. A full discussion of these subjects did not come within the limits of the writer's plan. The Military Art, in all its branches, is founded upon a comprehensive, and thorough knowledge of the exact and physical sciences; and in no one branch is the importance of this knowledge more felt, than in that of Engineering. So convinced of this truth was the Nation, which owes its high military standing, in no small degree, to its own Military Academies, and has itself served as a school to the world, that it bestowed on its Corps of Engineers, the remarkable title of *Corps du Génie*. In adding, therefore, these chapters, a clear comprehension of which supposes an advanced state of scientific attainment in the reader, it was far from the writer's intention, to lead his fellow citizens into error, by spreading a knowledge necessarily of a superficial kind amongst them; but he rather hopes, it may serve the purpose of fixing their attention more strongly and minutely upon a profession, which requires the season of youth to be spent, in thoroughly acquiring its elements, and the devotion of mature age to an incessant study of its principles, and their applications.

The author is indebted to Lieut. J. E. BLAKE, 6*th* INFANTRY, *for the very correct and handsome execution of the original drawings, from which the plates were engraved.*

TABLE OF CONTENTS.

CHAPTER I.

CHAPTER II.

CHAPTER III.

CHAPTER IV.

CHAPTER V.

CHAPTER VI.

CHAPTER VII.

CHAPTER VIII.

CHAPTER IX.

CHAPTER X.

CHAPTER XI.

CHAPTER XII.

CHAPTER XIII.

CHAPTER XIV.

FIELD FORTIFICATION.

CHAPTER I.

NOMENCLATURE AND GENERAL PRINCIPLES.

1. All dispositions made to enable an armed force to resist, with advantage, the attack of one superior to it in numbers, belong to the Art of Fortification.

2. The means resorted to, for the purpose of strengthening a position, may be either those presented by nature, as precipices, woods, rivers, &c., or those formed by art, as shelters of earth, stone, wood, &c.

3. If the materials used, are of a durable character, and the position is to be permanently occupied, the works by which it is strengthened, receive the name of *Permanent Fortification;*

but when the position is to be occupied only for a short period, or during the operations of a campaign, perishable materials, as earth and wood, are mostly used, and the works are denominated *Temporary* or *Field Fortification.*

4. The general appellation of *Intrenchments*, is applied to all field works; and a position strengthened by them, is said to be *Intrenched.*

5. To enable troops to fight with advantage, the intrenchments should shelter them from the enemy's fire; be an obstacle in themselves to the enemy's progress; and afford the assailed the means of using their weapons with effect. To satisfy these essential conditions, the component parts of every intrenchment should consist of a covering mass, or embankment, denominated the *parapet*, to shelter the assailed from the enemy's missiles, to enable them to use their weapons with effect, and to present an obstacle to the enemy's progress, and of a *ditch*, which, from its position and proximity to the parapet, subserves the double purpose of increasing the obstacle, which the enemy must surmount, before reaching the assailed, and of furnishing the earth to form the parapet.

6. From this view of the subject, intrenchments should be regarded only as accessaries in the defence of a position. They are inert masses, which, consuming a portion of the enemy's efforts, and detaining him in an exposed situation to the fire of the assailed, insure his defeat.

7. The general form of the parapet and ditch, is immediately derived from the preceding observations, and will be best understood by an examination of the *profile*, (Fig. 1,) which is a section of the intrenchment made, by a vertical plane, perpendicular to the general direction of the intrenchment.

8. The *exterior slope*, is the part of the parapet towards the enemy; it is always made with the same slope, that the earth when first thrown up, naturally takes.

The top of the parapet, denominated the *superior slope*, is the line along which the assailed fire on the enemy.

The *interior slope*, sometimes denominated the *breast height*, is the part against which the assailed naturally lean in the act of firing.

The *banquette* is the small terrace on which the soldier stands to deliver his fire; the top of it is denominated the *tread*, and the inclined plane by which it is ascended the *slope*.

The term *crest* is applied to those points of the profile, where a salient angle is formed; and where a re-entering angle is formed by two lines, the term *foot* is applied to the point, in connection with the name of the superior line; thus, *foot of the exterior slope; foot of the interior slope, &c.*

The *berm*, is the horizontal space left between the parapet and the ditch, to prevent the earth from yielding.

The slope of the ditch next to the parapet is the *scarp;* the opposite side the *counterscarp.*

The *glacis* is a small mound of earth raised in front of the ditch; it is seldom used in field works, and is therefore not a constituent part of their profile.

8. The profile shows only the form of an intrenchment in elevation, and by itself is not sufficient to show the relative bearing of all the

parts. A *plan*, or *trace*, which exhibits the direction of the different lines of the parapet, &c., is, in conjunction with the profile, requisite for this purpose.

The plan of intrenchments in general is so arranged as to procure a mutual defence of the parts. To effect this, certain parts are thrown forward towards the enemy, to receive his attack; they are denominated *advanced parts;* other portions, denominated *retired parts*, are withdrawn from the enemy, and protect by their fire the advanced parts. This arrangement naturally indicates that the general outline of the plan must present an angular system; some of the angular points, denominated *salients*, being towards the enemy, and others denominated *re-enterings*, towards the assailed.

10. When such a disposition is made, it is denominated a *flanked disposition;* because the enemy's flank is attained by the fire of the retired parts whilst he is advancing upon the salients.

11. A flanked disposition is shown in Fig 2. The advanced parts are denominated *faces*, the

retired parts, which protect the faces, are denominated *flanks;* the retired part connecting the flanks is the *curtain.*

The angle formed by the faces is denominated a *salient angle;* that formed by two retired parts a *re-entering angle;* and the one made by a face and the opposite flank an *angle of defence.*

The line bisecting a salient angle is denominated the *capital;* the distance from the salient to its opposite flank is the *line of defence.*

12. The form of the parapet, and the direction in which a soldier usually aims in firing over it, are the causes of two of the most important defects of intrenchments. Owing to the form of the parapet, and its height, the fire takes effect, only some distance beyond it, so that when the enemy has approached very near the parapet, particularly when he is in the ditch, the fire will pass over his head, unless the flanks are so arranged that their fire will sweep every point of the ditch; an arrangment of which particular angular systems are alone susceptible. This space, where the enemy can find a shelter, is, gene-

rally, in the ditches at the re-entering angles. It is denominated a *dead space*, or *dead angle.*

13. In delivering his fire a soldier usually aims directly to the front, so that the line of fire and the parapet make nearly a right angle with each other. In consequence of this the salients receive no protection from themselves, and there is an angular space in front of each of them, (which is equal to the supplement of the salient angle,) that is defended only by the fire of the flanks. This space is denominated a *sector without fire.*

The dead angles, therefore, depend partly on the flanking dispositions, but chiefly on the height of the parapet. The sectors without fire, are defects of the plan alone.

14. The attack and defence of intrenchments, bear a necessary relation to each other; and it is upon a knowledge of the course pursued by the assailant, that the principles regulating the defence should be founded.

15. An attack is, generally, opened by a fire of the enemy's artillery, whose object is to silence the fire of the intrenchments, and to drive the

assailed from the parapet; when this object is attained, a storming party, which usually consists of a detachment of engineer troops, a column of attack, and a reserve, is sent forward, under the fire of the artillery, to the assault. The detachment of engineer troops, precedes the column of attack, and removes all obstacles that obstruct its passage into the ditch. The line of march is directed upon a salient, through a sector without fire, and on the prolongation of the capital, as this line is least exposed to the fire of the works.

When the ditch is gained, the party makes its way to a re-entering angle, where, sheltered from the fire of the flanks, the work is entered by the column of attack; either by making a breach in the parapet, or else by means of ladders. The reserve supports the column of attack in case of need; and, if it is driven from the works, covers its retreat.

16. The manner of making the defence is with artillery, musketry, the bayonet and sorties. The enemy is attained at a distance, by the fire of the

artillery and musketry, whose effect will chiefly depend on the length of time that he is kept exposed to it by the ditch and the obstacles in front of it. The bayonet is resorted to, as soon as the enemy shows himself on the berm; and sorties are to be made, either when any irresolution or confusion, is seen in the enemy's ranks, or at the moment he is repulsed from the parapet.

It is from this general outline of the attack and defence, that the following principles, which regulate the plan and profile of intrenchments, are deduced.

I.

17. *A flanked disposition should be the basis of the plan of all intrenchments.*

The flanks sweep with their fire the ground in front of the faces; obviate sectors without fire and dead angles; cross their fire in front of the salients; and take the enemy's column in flank.

II.

Every angle of defence should be 90°.

An acute angle of defence exposes the faces to the fire of the flanks; an obtuse angle leaves a portion of the ground in front of the face undefended.

III.

A line of defence should not exceed 160 *yards.*

A close fire of musketry is more deadly than one of artillery, the musket will kill at distances between 250 and 300 yards, but its fire is not very certain beyond 160 yards; moreover, the enemy should be attained by the musketry before he gains the salients.

IV.

A salient angle should not be less than 60°.

A salient less than 60° is too weak to withstand the effects of frost, rain, &c.; the interior space which it encloses is too confined for the manœuvres of the troops; it forms a large sector without fire in front of it; the faces of acute salients are, from their position, more exposed to the enemy's enfilading fire than when the angle is obtuse.

V.

A strong profile is essential to a vigorous defence.

The rapidity with which a column of attack approaches, and the short time it remains exposed to the fire of the work, unless detained by obstacles in front of the ditch, render its loss, generally, so trifling as not to check its march, until it arrives at the crest of the counterscarp. Here, if the ditch is deep, some delay ensues in filling it up, or in cutting away the counterscarp, during which the column is exposed to a warm fire within short range. When the ditch is entered, a more serious obstacle remains to be encountered, in the additional height of the parapet and scarp; and when this obstacle is overcome, the enemy presents himself in a fatigued and exhausted state to the bayonets of the assailed, who have mounted on the top of their parapet to meet and drive him back into the ditch.'

VI.

The bayonet should be chiefly relied on to repel the enemy.

Unless the assailed are determined to meet the enemy at the point of the bayonet, they must evacuate their works so soon as he has entered the ditch; a longer delay of an attempt to retreat, would be followed by the most disastrous consequences.

The results of innumerable actions prove that the defence with the bayonet is the surest method of repelling the enemy. The assailed, having now become the assailant, are assisted by that moral effect which is produced by a change from a defensive to an offensive attitude. They have moreover the advantages of position and freshness over a climbing and exhausted enemy.

VII.

Intrenchments should be arranged to facilitate sorties.

A sortie made on the flank of the enemy, at the moment when his column is either checked by the fire of the works, or is in a state of disorder when entering the ditch, will generally prove decisive in repelling his attack.

This principle, however, is applicable only to engagements with large bodies of troops, defending works open in their rear. A small detachment should rest satisfied with repulsing the attack; and should not give up the advantages of their position, by sallying out to engage the enemy on equal terms.

VIII.

Intrenchments should contain a reserve proportioned to their importance.

The troops engaged in the immediate defence of the works, might be overpowered by reiterated assaults of the enemy, unless they were supported by a reserve. The duties of the reserve are to charge the enemy in any critical moment of disorder, and to cover the retreat of the troops if driven from the parapet.

IX.

Intrenchments should be defended to the last extremity.

The chief object of intrenchments is to enable the assailed to meet the enemy with success; by first compelling him to approach, under every disadvantage of position, and then, when he has been cut up, by the fire of the works, and is exhausted by his efforts to reach the parapet, to assume the offensive, and drive him back at the point of the bayonet. This object can only be attained by defending the works to the last extremity; and unless attained, intrenchments would serve no other purpose than to shelter the assailed from the enemy's fire; for the damage received by the enemy, from the fire of troops, who see safety only in retreat, and not in their own courageous efforts to repel the assault, will be necessarily trifling.

This principle leads to the rejection of all advanced works thrown up in front of the principal intrenchments; and to all dispositions of works in several lines, where the object is to retreat, successively, from one to the other. If an advanced work is required for the defence of a point, which cannot be defended by the principal works, it should be supported by a reserve, and its garrison not be suffered to retreat, until it is in danger of being overpowered: when it should retire under cover of the reserve. As to retreating from one line to another, one of two things must happen; either the first line must be evacuated, before the enemy enters the ditch, in order that the assailed may gain their second line in safety; in which case the first line will be of little service: or else, if the assailed wait until the assault is made on the parapet before retreating, the enemy will enter pell-mell with them into the second line, which will thus be of no service.

Troops in action cannot enter into the spirit of fortification on paper. A retreat carries with it all the moral effects of a defeat; it inspirits the assailant; renders the retreating corps timid; and impairs the confidence of the troops of the second line, in the strength of their own position, when they see the first line carried with such ease. Add to this, the confusion that must ensue among the best disciplined troops, under such circumstances, and the importance attached to the principle will be fully justified.

CHAPTER II.

PLAN OF SIMPLE INTRENCHMENTS OR THE ELEMENTS OF LINES.

18. The ground occupied by a work is denominated the *site*, or *plane of site*.

The *command* is the height of the interior crest above the site; and the *relief* is the height of the same line above the bottom of the ditch.

19. A fire is said to be *direct*, *slant*, or *enfilading*, according as its direction is perpendicular to, makes an angle of 30° with, or is on the prolongation of the line at which it is aimed; when the line is taken in the rear the fire is denominated a *reverse fire*; and when a given space is defended by the fire from several points crossing over it, the defence is denominated a *cross fire*.

20. In planning a work the interior crest is regarded as the directing line in regulating the dimensions of the faces, flanks, &c, because this

line shows the column of fire for the defence. There exists a necessary subordination between the plan, relief, and command of works, which prevents the dimensions of the one being regulated independently of the others; but, without entering into a close examination of this necessary co-relation of the parts, it may be stated generally, that faces should vary between 30 and 80 yards, flanks between 20 and 40 yards, and curtains should not be less than twelve times the relief.

21. A great variety of figures has been used for the plan of simple intrenchments. They may all be reduced to the following, the *Right Line;* the *Redan;* the *Lunette*, or *Detached Bastion;* the *Crèmaillére*, or *Intended Line;* the *Priest-Cap*, or *Swallow-Tail;* the *Redoubt;* the *Star Fort;* and the *Bastion Fort.*

22. The redan (Fig. 3) is a work consisting of two faces; the *gorge*, or entrance in the rear, being open. This work is used to cover a point in its rear; such as a bridge, defile, ford, &c. Having no flank defences its salient is unprotec-

ted, and to obtain a fire in the direction of its capital a short face, denominated a *pan coupé*, is sometimes made in its salient angle.

23. The lunette (Fig. 4) consists of two faces and two flanks. This work is used for the same purposes as the preceding. It has the same defects; but possesses the advantage of sweeping with the fire of its flanks ground which might be badly defended by its faces.

24. The indented line (Fig. 5) serves to convert the direct fire of a right line into a flank and cross fire, and is therefore frequently subtituted for the right line.

25. The priest-cap (Fig. 6) is seldom used as a detached work; but is generally combined with the right line and the indented line to procure a flank, or a cross fire, in front of them.

26. Any enclosed work of a polygonal form, without re-entering angles, is denominated a redoubt. This work is used to fortify a position which can be attacked on all sides; the works which have already been described, being unsuitable for this purpose, as their gorges are open,

and therefore require to be supported by troops, or works, in their rear; except when they are so situated that an attack cannot be made at the gorge.

The square (Fig. 7) is the most common form for a redoubt, on account of the ease with which it is constructed, and the advantage it possesses, when combined with several others, of protecting the spaces between them by a cross fire.

All redoubts have the same defects. The ditches are unprotected, and there is a sector without fire in front of each salient. For the purpose of remedying the sector without fire, it has been proposed to convert a portion of each face at the salient angles, into an indented line, to procure a fire in the direction of the capitals. This method is not of practical application; and if it could be applied would only serve the purpose of changing the position of the sectors without fire from the salients to other points.

27. The star fort takes its name from the form of the polygonal figure of its plan. (Figs. 8, 9, 10, 11.) It is an enclosed work, with salient and

EXPLANATIONS OF PLATE I.

Fig. 1. This Fig. shows the ordinary form of the Profile of an Intrenchment, in soils of which the Natural Slope is one perpendicular to one base.

A B C D E F is the profile of the Parapet.
G H I K the profile of the Ditch.
L M N the profile of the Glacis.
A B is the Banquette Slopes.
B C Tread of the Banquette.
C D the Interior Slope.
D E the Superior Slope.
E F the Exterior Slope.
F G the Berm.
G H the Scarp.
H I the Bottom of the Ditch.
I K the Counterscarp.
A the Foot of the Banquette Slope.
B the Crest of the Banquette.
C the Foot of the Interior Slope.
D the Interior Crest.
E the Exterior Crest.
F the Foot of the Exterior Slope.
G the Crest of the Scarp.
H the Foot of the Scarp.
I the Foot of the Counterscarp.
K the Crest of the Counterscarp.
M the Crest of the Glacis.
N the Foot of the Glacis.

Fig. 2. This Fig. shows the general Plan of Intrenchments with Flanking arrangements.

O A B and E F P are the Advanced Parts.
B C D E are the Retired Parts.
A O, A B, E F, and F P are the Faces.
B C and D E are the Flanks.
C D the Curtain.
A D and C F the Lines of Defence.
O A B and E F P are the Salient Angles.
B C D and C D E the Re-entering Angles.
A D F and B C F the Angles of Defence.
b A c and e F g the Sectors without Fire.
L M and N R the Capitals.

Fig. 3. This Fig. shows the Plan of the Redan.

A B and C D are the Faces.
A D is the Gorge, or Gorge Line.
B C is a Pan Coupé.

Fig. 4. Shows the plan of the Lunette.

B C and C D are the faces.
A B and E D the Flanks.

Fig. 5. Shows the plan of the Crémaillère, or Indented Line.

Fig. 6. Shows the plan of the Priest-cap, or Swallow-tail

Fig. 7. Shows the plan of the Square Redoubt, with the faces at one angle indented, and at the other arranged with a pan-coupé.

Figs. 8, 9, 10 and 11. Show plans of Star Forts for polygons of four, five, six, and eight sides.

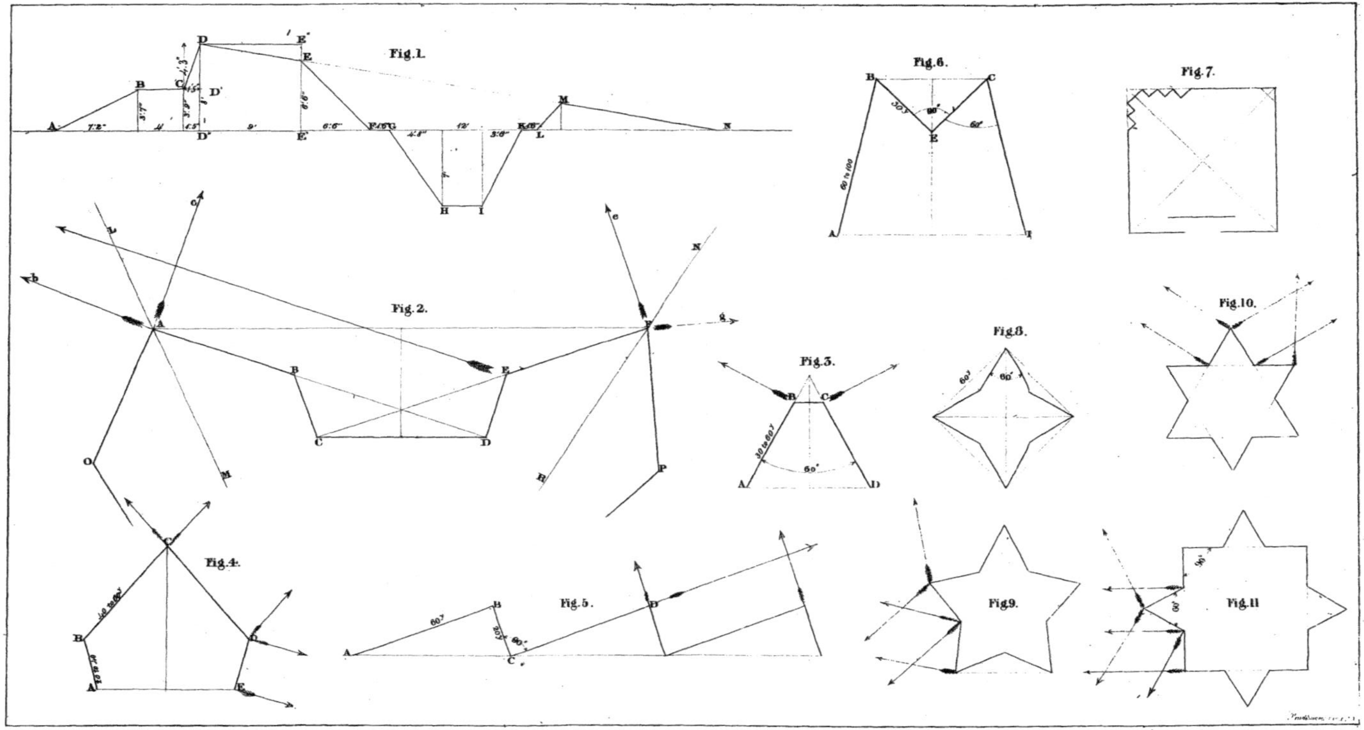

Fig. 1.
Fig. 2.
Fig. 3.
Fig. 4.
Fig. 5.
Fig. 6.
Fig. 7.
Fig. 8.
Fig. 9.
Fig. 10.
Fig. 11.

re-entering angles; the object of this arrangement being to remedy the defects observed in redoubts. This, however, is only partially effected in the star fort: for, if the polygon is a regular figure, it will be found, that, except in the case of a fort with eight salients, the fire of the faces do not protect the salients; and that in all cases there are dead angles at all the re-enterings. The star fort has moreover, the essential defect, that, occupying the same space as a redoubt, its interior capacity will be much less, and the length of its interior crest much greater, than in the redoubt: it will, therefore, require more men than the redoubt for its defence, whilst the interior space required for their accommodation, is diminished. These defects, together with the time and labor required to throw up such a work, have led engineers to proscribe it, except in cases where they are compelled by the nature of the site to resort to it.

To plan a star fort, its salients should not be less than 60°, and its faces may vary from 30 to 60 yards.

28. The bastion fort, satisfies more fully the conditions of a good defence, than any other work; but, owing to the time and labor required for its construction, it should be applied only to sites of great importance, which demand the presence of troops during the operations of a campaign.

The bastion fort may consist of a polygon of any number of sides; but for field forts, the square and pentagon are generally preferred, owing to the labor and construction. To plan a work of this kind, a square or pentagon (Fig. 12) is laid out, and the sides bisected by perpendiculars; a distance of one eighth of the side is set off on the perpendiculars in the square, or one seventh in the pentagon; from the angular points of the polygon, lines are drawn through the points thus set off; these lines give the direction of the lines of defence; from the salients of the polygon distances, equal to two sevenths of the side, are set off on the direction of the line of defence, which give the faces; from the extremities of the faces, the flanks are drawn perpendicular to the lines of defence;

the extremities of the flanks are connected by the curtains.

The side of the polygon is termed the *exterior side;* the line bisecting it, the *perpendicular;* the angle at the salient, is the *flanked angle;* the one formed by a face and flank, the *shoulder angle;* the one between the flank and curtain, the *angle of the curtain;* the portion of the work included between the capitals of two adjacent bastions is denominated a *bastioned front,* or simply a *front.*

An examination of the arrangement of a bastioned front will show that there are neither dead angles nor sectors without fire; that the salients, and all the ground within the range of musketry, are protected by formidable columns of direct, flank, and cross fire. There is one point in this system that demands particular attention, which is, that the counterscarp of the ditch, if laid out parallel to the interior crest, would form a dead angle along each face near the shoulder; because the fire of the flank would be intercepted by the crest of the counterscarp. To prevent this, either the counterscarps of the faces must be produced to intersect, and all the earth between them and the scarp of the flanks and curtain, be excavated, (Fig. 13,) or the ditch of each face must be produced in a slope from the bottom, opposite the shoulder, (Fig. 14,) so that it can be swept by the fire of the flank.

The first method is the best, but requires most labor; the second is chiefly objectionable as it gives an easy access to the ditch, which might be taken advantage of in an assault. It is proposed, to obviate this, to dig a second ditch at the foot of the slope across the main ditch, twelve feet wide, and about six feet deep; to make it pointed at the bottom, and to plant a row of palisades in it. The profile through the flank (Fig. 14) shows this arrangement.

Forts have been proposed with half bastions, but, being very little superior to the redoubt, and much more difficult of construction, they ought never to be used.

The exterior sides of the bastion fort should not exceed 250 yards, nor be less than 125 yards, otherwise the flanking arrangements will be imperfect. With a relief of 24 feet, which is the greatest that, in most cases, can be given to field works, and an exterior side of 250 yards, the ditch of the curtain will be perfectly swept by the fire of the flanks, the lines of defence will be nearly 180 yards, a length which admits of a good defence, and the flanks will be nearly 30 yards. With a relief of 14 feet, the least that will present a tolerable obstacle to an assault, and an exterior side of 125 yards, the ditch of the curtain will be well flanked, the flanks will be nearly 20 yards in length, and the faces between 30 and 40 yards. Between these limits, the dimensions of the exterior side must vary with the relief.

29. The defence of enclosed works demands that every point of the parapet should be guarded, at the moment of assault, either by cannon or musketry. The troops may be drawn up for the defence either in one, two, or three ranks; and there should, moreover, be a reserve proportioned to the importance attached to the work. The

free interior space, denominated the *terre-plein*, should be sufficiently great to lodge the troops, with the cannon and its accessaries, and will therefore depend on the nature of the defence. The following data will suffice to regulate this point.

Each man will occupy one yard, linear measure, along the interior crest, and each cannon from five to six yards. The space requisite to lodge each man is one and a half square yards; and about sixty square yards should be allowed for each gun. Besides this space an allowance must be made for the *traverses*, which are mounds of earth thrown up in the work to cover an outlet, to screen the troops from a reverse, or an enfilading fire, &c.; and for powder magazines, when they are not made in the traverses. The area occupied by a traverse will depend on the dimensions of the traverse, and cannot be fixed beforehand; that allowed for a magazine for three or four cannon may be estimated at fifteen or twenty square yards.

30. As a field fort must rely entirely on its own

strength, it should be constructed with such care that the enemy will be forced to abandon an attempt to storm it, and be obliged to resort to the method of regular approaches used in the attack of permanent works. To effect this, all the ground around the fort, within the range of cannon, should offer no shelter to the enemy from its fire ; the ditches should be flanked throughout ; and the relief be so great as to preclude any attempt at scaling the work.

CHAPTER III.

DISCUSSION OF THE GENERAL PROFILE.

31. The general form of the parapet is the same for all works. Its dimensions will vary with the kind of soil used in its construction; with the time and means that can be employed; with the time that the work is to remain occupied; and, finally, with the time and means that the enemy can dispose of in the attack, and the degree of resistance that the work should offer.

32. The command of the interior crest (Fig. 1) should be regulated so as to intercept the enemy's missiles, and to shelter the assailed. Men of the greatest ordinary stature, in bringing their muskets to an aim, do not fire at a higher level than about five feet; therefore any mass of this height in front of them will just intercept their fire; but this mass would not shelter a man standing

behind it; to effect this, in the case of the tallest men usually found in the ranks, the interior crest should be at least six feet six inches above the terre-plein. The command must then be regulated by these two facts, and this principle may be established. *The command of a field work over the ground occupied by the enemy must never be less than five feet; nor less than six feet six inches over that occupied by the assailed.*

But this minimum command would give the assailed but a slight advantage, as the men, when on the banquette, would be very much exposed; and in an assault the height of the parapet would present an inconsiderable obstacle. These defects have led engineers to adopt eight feet as the least height of parapet which will admit of a respectable defence. The greatest height has been fixed at twelve feet, owing to the difficulty of throwing up a work with the ordinary means at hand, which are usually only the pick and shovel.

The thickness of the parapet, which is always estimated by the horizontal distance, between the interior and exterior crests, is regulated by the material used for the parapet; the kind of attack; and its probable duration.

Shot will penetrate ordinary earth, when well rammed, the distances laid down in the following table:

Musket ball,	1 foot 6 inches
6 pound shot,	$3\frac{1}{2}$ to $4\frac{1}{2}$ feet
9 do.	$6\frac{1}{2}$ to 7 do.
12 do.	$8\frac{1}{2}$ to 10 do.
18 and 24 do.	$11\frac{1}{2}$ to 13 do.

In order to insure the safety of the troops, these dimensions are augmented one half; so that no shot shall penetrate more than two-thirds the entire thickness.

34. The superior slope is arranged to defend the crest of the counterscarp; to effect which the fire should not strike below the crest, nor pass more than three feet over it; otherwise, either the counterscarp would be damaged, or the assailed by stooping when near the crest would find a shelter. The inclination of the superior slope, however, should not be greater than one fourth,

nor less than one sixth, that is, the base of the slope should be between four and six times the height. If greater than one fourth, it would make the portion of the parapet, about the interior crest, too weak; and if less than one sixth, the ground in front of the work, would not be so well defended; moreover, artillery cannot be fired at a greater depression than one sixth, without injuring the carriage; and this inclination of the superior slope serves as a check in rapid firing.

If, owing to the command, the fire should pass higher than three feet above the crest of the counterscarp, it would then be necessary to construct a glacis in front of the ditch. It must be arranged in such a manner, that it can be swept by the fire of the work; and be commanded at least five feet by the work.

35. The exterior slope is the same that the earth naturally assumes. Any means used to make it steeper would be injurious; because they would be soon destroyed by the enemy's fire, and the earth giving way, would diminish the required thickness of the parapet.

36. The interior slope receives a base equal to one-third its height. This is a result of experience, which has shown that it is the most convenient for the soldier in leaning forward to deliver his fire over a parapet.

37. The tread of the banquette is placed four feet three inches below the interior crest; this will admit men of the lowest ordinary stature, to fire conveniently over the parapet. Its width is two feet for a defence with one rank; and four feet for two or three ranks; because the third rank does not fire, and is therefore placed on the banquette slope, the base of which is twice the altitude, to render the ascent convenient. When the tread of the banquette is very high, and particularly in enclosed works, where interior space is wanted, steps may be substituted for a slope; the rise of the step should be nine inches, and its breadth twelve inches. The tread of the banquette should receive a slope of two inches to the rear to drain off the surface water.

38. The berm is a defect in field works, because it yields the enemy a foot-hold to breathe

a moment, before attempting to ascend the exterior slope. It is useful in the construction of the work for the workmen to stand on; and it throws the weight of the parapet, back from the scarp, which might be crushed out by this weight. In firm soils, the berm may be only from eighteen inches to two feet wide; in other cases, as in marshy soils, it may require a width of six feet. In all cases, it should be six feet below the exterior crest; to prevent the enemy, should he form on it, from firing on the troops on the banquette.

39. The ditch should be regulated to furnish the earth for the parapet. To determine its dimensions, the following points require attention; its depth should not be less than six feet, and its width less than twelve feet, to present a respectable obstacle to the enemy. It cannot, with convenience, be made deeper than twelve feet; and its greatest width is regulated by the inclination of the superior slope, which, produced, should not pass below the crest of the counterscarp.

40. The slopes of the scarp and counterscarp will depend on the nature of the soil, and the

action on it of frost and rain. The scarp is less steep than the counterscarp; because it has to sustain the weight of the parapet. It is usual to give the slope of the scarp, a base equal to two-thirds of the base of the natural slope of a mound of fresh earth whose altitude is equal to the depth of the ditch; the base of the counterscarp slope is made equal to one half the same base.

41. To determine the exact dimensions of the ditch, for a given parapet, requires a mathematical calculation, which will be given in a Note. On the field a result may be obtained, approximating sufficiently near the truth for practice, by assuming the depth of the ditch and dividing the surface of the profile of the parapet by it to obtain the width. In excavating the ditch it will be found that more earth will be furnished at the salients than is required for the parapet; and that the re-enterings will not always furnish enough. On this account, the width of the ditch should not be uniform, but narrower at the salients than the re-enterings.

CHAPTER IV.

MANNER OF REGULATING THE RELIEF OF INTRENCHMENTS ON IRREGULAR SITES.

42. When a work is placed on level ground, it usually receives a uniform relief; but when the site is irregular, or there are commanding eminences within cannon range, a uniformity of relief cannot be preserved, because the interior of the work might be exposed to the enemy's view, from the commanding points.

43. The plan will also be modified by the same causes. The principal faces should be placed so as not only to guard all the points where an enemy might approach; but the enemy should not be able to take up their prolongations to obtain an enfilading, or a reverse fire upon them. The position of the points to be guarded, and that of the commanding eminences, require to be carefully studied, before adopting any definitive plan.

The only general rules that can be laid down, are to place the principal lines so as to obtain a direct and cross fire, on the approaches of the enemy; and placing them, at the same time, as nearly parallel as practicable, to the general direction of the crests of the commanding heights, so that the enemy occupying the crest will have a direct fire alone on them.

44. When the enemy occupies a position more elevated than the work, he is said to have a *plunging fire* on it; and when the relief of the work is regulated so as to intercept this fire, the work is said to be *defiladed*.

43. The *defiladement* of field works is not indispensable to a good defence; nor is it generally practicable, except in small enclosed works. It is however, not only a conservative means which should be applied when practicable; but it also inspires the assailed with more confidence; for the soldier regards with distrust the strength of his position, when he finds himself exposed to the view of the enemy, from an elevated point.

44. The defiladement of a work, is a practical operation performed on the ground in the following manner:—

Let A B C D E (Fig. 15) be the plan of a work, a lunette for example; and the points O, O, &c., the most elevated points of a commanding position in front of the work. At the points A, B, &c., let straight poles be planted vertically, and and on the poles along the gorge line let a point be marked, at three feet above the ground. Let two pickets be driven in the ground along the gorge line, and a cord *a*, *b*, or a straight edge of pine, be fastened to them, on the same level as the two points marked on the poles at A and E. Let an observer now place himself in the rear of *a b*, so as to bring the poles at B, C, and D, and the points O, O, &c., within the same field of vision. Let observers be placed at B, C, and D. The first observer now sights along *a b*, until he brings his eye in the position that *a b* will appear tangent to the most elevated of the points O. Having accurately determined this position, he next directs the other observers to slide their hands along the poles until they are brought into the same plane of vision with the point o, and the line *a b*, and to mark those points on the poles. These points, together with the two first marked, will evidently be in the same plane, and this plane, produced, will be tangent to the highest point o. It is denominated the *Rampant Plane.* Now if a point be marked on each pole, at five feet above the points thus determined, these points will be contained in a second ideal plane, parallel to the first, and five feet above it. This plane is denominated the *Plane of Defiladement*, and the interior crests of the work are contained in this plane, being the lines joining the highest points marked on the poles.

As the gorge line is farthest from the heights, and the rampant plane ascends towards them, it will necessarily pass at more than three feet above every other point of the terre-plein of the work; and the plane of the defiladement, in like manner, will pass at eight feet above the terre-plein at the gorge, and at five feet above the highest point o. A plane of defiladement is therefore defined to be, *that plane which, containing the interior crests of a work, passes at least at eight feet above every point of the terre-plein, and at least five feet above every point that the enemy can occupy within the range of cannon,* which range may be taken, with safety, at one thousand yards.

45. When a work is placed in a hollow formed by two eminences, and is exposed to both a direct and reverse fire from them, it cannot be defiladed, by the means just explained, without giving it a relief altogether too great for field works. To avoid this the method of reverse defiladement must be resorted to.

46. Let A B C D E (Fig. 16) be the plan of a lunette, placed between the two eminences o, o′; its face and flank A B C being exposed to the direct fire from o, and a reverse fire from o′; the other face and flank being in like manner exposed to a direct and a reverse fire.

47. Suppose a section of the work (Fig. 17) to be made by a vertical plane, passing through the

highest points o and o′. If in this plane a vertical *a b*, be drawn, corresponding to the capital of the work, and eight feet be set off on this vertical from the point *a*, and two verticals be drawn through the points o and o′, and five feet be set off on each of them; and then the points *c* and *c′* be joined with *d*, it is obvious that the interior crest of the parapet A, being placed on the line *c d*, will screen all the ground in the rear of it, as far as the capital, from the direct fire from o. The parapet B being regulated in a similar manner, will screen all the ground behind it as far as the same line. But the fire from o′ would take the parapet A in reverse, and that from o the parapet B; to prevent this, a covering mass, denominated a *traverse*, must be erected on the line of the capital, and a sufficient heighth be given to it to screen both A and B from a reverse fire. To effect this, let eighteen inches be set off on a vertical through the interior crests of A and B; the point *e* being joined with *c′*, and the point *e′* with *c*; it is here also obvious, that if the top of the traverse be placed on the line *c e′*, it will effectually screen both the parapets from all reverse fire; because every shot that strikes the top of it will pass at least eighteen inches above the two parapets, and, since the banquettes are four feet three inches below the interior crests, the shot must pass five feet nine inches above the banquettes, which will be quite sufficient to clear the heads of the men when on the banquettes. This illustration explains the spirit of the method of reverse defiladement. The operation itself is performed in the following manner:

48. Poles (Fig. 16) are planted at the points A B C, &c., and one at the point F, where the lines of the capital and gorge intersect. On the pole F, a point is marked three feet above the ground, and a point is likewise marked on the pole at C, which should be one foot six inches higher than that on F; that is, if the ground between the two poles be level, the point on C will be four feet six inches above the ground. Two stout pickets may next be planted between F and C, and a cord, or a straight edge, be fastened to them, so as to be

in the same line as the points marked on the poles.

Observers are then placed at the poles A and B; and another places himself behind the cord, so as to bring the points O, A, and B, in the field of vision with it; then shifting the position of the eye until the cord is brought tangent to the highest point on O, he directs the observers at A, and B, to mark on the respective poles the points where the plane of vision intersects them. This operation will determine the rampant plane for one half the work A B C F, that for the other half will be determined by a similar process. If then a distance of five feet be set off on each pole above the points thus determined, these points will fix the position of the interior crests.

It is obvious that the interior crest of the part A B C is not in the same plane as that of the part C D E. These two planes are denominated *planes of direct defiladement.*

49. To determine the height of the traverse is the next step. To do this, the height of the tread of the banquette is ascertained on the three poles,

B, C, D, and a distance of nine inches is set off on each pole above the tread. Between the points thus determined a cord is stretched, or if the distance be too great for this, two pickets might be placed between B, and C, and a cord, or straight edge, be fastened to them in the required direction. An observer is then placed at the pole F, and another places himself behind the line B C, so as to bring the cord, and the points O′, and F, in the field of vision; he then shifts the position of the eye until the cord is brought to touch the point O′; he then directs the observer at F, to mark the point on the pole where it is intersected by the plane of vision. A similar operation is performed with the point O, and the face C D, and above the highest point thus determined on F, a distance of five feet is set off for the top of the traverse at F; and five feet nine inches is set off above the tread of the banquette at C for the top of the traverse at that point.

The planes which determine the top of the traverse, are termed *planes of reverse defilade-ment.*

50. The traverse is finished on top like the roof of a house, with a slight pitch; its thickness at top should seldom exceed ten feet, and will be regulated by the means the enemy can bring to the attack; its sides are made with the natural slope of the earth; but, when the height of the traverse is considerable, the base of the side slopes would occupy a large portion of the interior space; to remedy this, in some measure, the portion of the sides which are below the planes of direct defiladement, may be made steeper than the natural slope; the earth being retained by a revetment of sods, &c.

When the salient of the work is arranged for defence, the traverse cannot be extended to the salient angle; it is usual to change its direction within some yards of the salient, and unite it with the face most exposed. Traverses are also used to cover faces exposed to an enfilade fire; for this purpose they are placed perpendicular to the face to be covered. If several are required, they may be placed twenty or thirty yards apart; each traverse should be about twenty-four feet long, and thick enough to be cannon proof.

52. The cases of defiladement here examined, are those of works open at the gorge; the same principles, and similar methods, would be applied to enclosed works. After the plan of the work has been regulated, the arrangement of the traverses next demand the attention; the only rule that can be laid down is, to place them in the most favorable position to intercept the reverse and enfilading fire of the enemy; and if there should be a choice with respect to several positions, to select the one which will give the lowest traverse.

53. The difficulty of defiladement, owing to the great relief of the parapets, the labor of erecting the traverses, and the room which they occupy within the work, which is frequently wanted for the defence, restricts the employment of it to enclosed works, which are to remain occupied during some time, and whose position, owing to some point to be defended, cannot be shifted.

54. If it is not probable that a commanding eminence should be occupied by the enemy, nevertheless should the defence be not impaired, it would be best to place the work beyond the cannon range of the eminence.

55. The irregularity in the profile of the parapet, caused by defiladement, will occasion a correspondent irregularity in the profile of the ditch. Where the parapet is highest, the ditch will require to be widest and deepest; for, in order to avoid the removal of the earth to considerable distances, it is best that the earth for each portion of the parapet should be taken from the ditch in front of it. No other rule can be laid down in this case, than to keep the dimensions of the ditch within the prescribed limits; and, if this will not admit of its counterscarp being well defended, to raise a glacis in front of it, subject to the fire of the work.

CHAPTER V.

MANNER OF THROWING UP A WORK.

56. THE foregoing chapters contain all that is requisite to determine the plan and relief of field works under all circumstances of variety of ground. To follow a natural order, the next step will be to describe the manner of laying the work out on the field, which is termed *profiling;* the distribution of the workmen to excavate the ditch, and form the parapet; and the precautions to be observed in the construction.

57. Poles (Fig. 18) having been planted at the angles of the work, and the height of the interior crest marked on them, a line is traced on the ground, with a pick, showing the direction of the interior crests. At suitable distances, say from twenty to thirty yards apart, cords are stretched between two stout pickets, in a direction perpendicular to the line marked out by the pick;

these cords should be exactly horizontal. A stout square picket is driven firmly into the ground, where the cord crosses above the pick-line, and a slip of pine, on which the height of the interior crest is marked, is nailed to the picket. The thickness of the parapet is measured on the cord, and a picket driven into the ground to mark the point. The base of the interior slope, and the tread of the banquette, are set off in a similar manner; and a slip of deal is nailed to each of the pickets. The height of the exterior crest, and the tread of the banquette, are easily ascertained, from the position of the cord, and the interior crest; these points having been marked on their respective slips, the outline of the parapet is shown by connecting them by other slips, which are nailed to the uprights; the banquette slope, and exterior slope, will be determined by a similar process.

58. From the profiles thus formed perpendicular to the interior crests, the oblique profiles at the angles can readily be set up, by a process which will suggest itself without explanation.

EXPLANATIONS OF PLATE II.

Fig. 12 Shows the Plan of a Bastioned Fort, constructed on a Square.

A B is the Exterior Side.

C D the Perpendicular.

A the Salient Angle.

E the Shoulder Angle.

H the Angle of the Curtain.

Note. The nomenclature of the other parts is the same as in Fig. 2.

Fig. 13. Shows the plan of a Bastioned Front, with the entire ditch excavated in front of the curtain C.

Fig. 14. Shows the same with the ditches of the faces prolonged upwards to C in a slope. This arrangement will be best understood from the Section Fig. 14.

Fig. 15. A B C D E is the position of the interior crest marked out on the ground; a e is a cord or straight edge, placed along the gorge line A E, three feet above the ground; o, o′ are the commanding eminences.

Fig. 16. F C is the position of the capital, with a cord stretched between the poles at F, C, the other lines same as Fig. 15.

Fig. 17. A and B are the profiles of the parapets; C that of the traverse; o, o′ the commanding eminences; c d, c′ d′ the lines of direct fire; c e and c′ e′ the lines of reverse fire.

Fig. 18. Shows a profile made of slips of pine or of cord.

Plate 2.

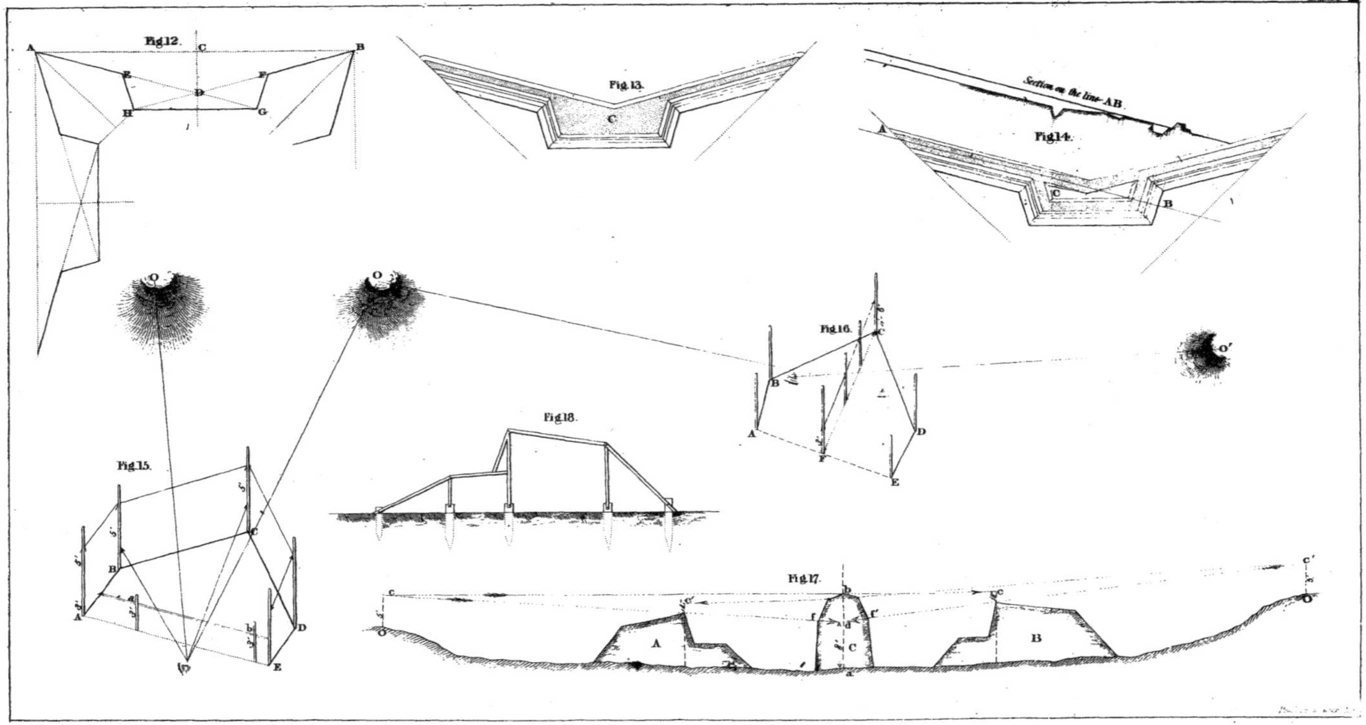

Fig.12.
Fig.13.
Fig.14.
Section on the line AB.
Fig.15.
Fig.16.
Fig.17.
Fig.18.

59. Having completed the profiling, the foot of the banquette, and that of the exterior slope, are marked out with the pick, and also the crests of the scarp and counterscarp. All the arrangements preparatory to commencing the excavation are now complete.

60. Experience has shown that, in ordinary soils, a man with a pick can furnish employment to two men with shovels; that, not to be in each other's way, the men should be from four-and-a-half to six feet apart; and, finally, that a shovel full of earth can be pitched by a man twelve feet in a horizontal direction, or six feet in a vertical direction.

61. To distribute the workmen, the counterscarp crest is divided off into lengths of twelve feet, and the interior crest into lengths of nine feet. These points might be marked out by pickets numbered one, two, three, &c. In each area, thus marked out, a working party is arranged consisting of a pick with two shovels placed near the counterscarp, two shovels near the scarp, and one man to spread, and one to ram the earth, for two working parties.

62. The pick commences by breaking ground so far from the counterscarp crest (Fig. 19) that, by digging vertically three feet, he will arrive at the position of the counterscarp. The excavation is carried on by digging to the depth of three feet, advancing towards the scarp, where the same precaution is observed as at the counterscarp. The earth is thrown forward, and evenly spread and rammed, in layers of about twelve inches, from the banquette slope to the exterior slope.

63. For the facility of entering the ditch, the offsets, at the scarp and counterscarp, may be formed into steps with a rise of eighteen inches each; and, if the ditch is deeper than six feet, an offset, about four feet broad, should be left at the scarp, about mid-depth of the ditch, to place a relay of shovels to throw the earth on the berm. In some cases, a scaffold of plank is raised in the ditch for the same purpose.

64. When the ditch has been excavated to the bottom, the offsets are cut away, and the proper slopes given to the sides. The earth furnished by

the offsets, if not required to complete the parapet, may be formed into a small glacis.

65. If the soil is stony, the vegetable mould on the surface should be scraped off, and reserved to form the top of the parapet, which should be made of earth of this kind, to the depth of at least eighteen inches, to prevent injury to the troops from the effect of a shot striking the top, and scattering the pebbles in their faces.

66. In making the parapet, care should be taken to form a drain, at some suitable point, to carry off the water from the interior into the ditch. The water from the drain should not be suffered to run down the scarp, as it would soon destroy it. A gutter, formed of boards, should be made to prevent this.

67. The time required to throw up a work will depend on the nature of the soil and the expertness of the laborers. From troops unaccustomed to the use of ditching tools, six cubic yards may be considered a fair day's work in ordinary soils, when the earth is not thrown higher than six feet; but when a relay is placed on an offset in the

ditch, from four to five cubic yards may be taken as the result of a day's work for each man. Expert workmen will throw up from eight to ten cubic yards at task-work.

CHAPTER VI.

REVETMENTS.

68. A *revetment* consists of a facing of stone, wood, sods, or any other material, to sustain an embankment, when it receives a slope steeper than the natural slope.

69. In field works revetments are used only for the interior slope of the parapet and for the scarp; for the first sods, pisa, fascines, hurdles, gabions and plank, are chiefly used; and for the last, timber.

70. *Revetment of sods.* Sod work forms a strong and durable revetment. The sods should be cut from a well-clothed sward, with the grass of a fine short blade, and thickly matted roots. If the grass is long, it should be mowed before the sod is cut.

Sods are of two sizes, (Fig. 20,) one termed *stretchers*, are twelve inches square, and four-and-a-half inches thick; the others, termed *headers*, are eighteen inches long, twelve inches broad, and four-and-a-half inches thick.

71. The sod revetment (Fig. 21) is commenced as soon as the parapet is raised to the level of the tread of the banquette. A layer of sods is then laid, either horizontal or a little inclined from the banquette; the layer consists of two stretchers and one header alternating, the end of the header laid to the front. The grass side is laid downward; and the sods should protrude a little beyond the line of the interior slope, for the purpose of trimming the layer even at top, before laying another, and to make the interior slope regular. The layer is firmly settled, by tapping each sod as it is laid with a spade or a wooden mallet; and the earth of the parapet is packed closely behind the layer.

A second layer is laid on the first, so as to cover the joints, or, as it is termed, to *break joints* with it; using otherwise the same precautions as with the first. The top layer is laid with the grass up; and in some cases pegs are driven through the sods of two layers to connect the whole more firmly, which is, however, by no means necessary to form a strong sodding.

72. When cut from a wet soil, the sods should not be laid until they are partially dried; otherwise they will shrink, and the revetment will crack in drying. In hot weather the revetment should be watered frequently, until the grass puts forth. The sods are cut rather larger than required for use; and are trimmed to a proper size from a model sod.

73. *Pisa revetment.* Ordinary earth, if mixed with a proper proportion of clay, and the whole be well kneaded with just water enough to cause the particles to adhere when squeezed in the hand, may be used for a revetment, and is termed a pisa revetment. Sometimes chopped straw is mixed up with the mass to cause it to bind better.

74. The pisa is laid in layers of twelve inches thick, and two feet broad, and well packed. The same precautions should be taken in forming the parapet behind it as in sod revetments. The face of the revetment may be sowed with grass seed

or oats, and when the stalk comes to maturity it should not be cut, but suffered to remain as a kind of thatch to protect the facing from the weather.

75. *Fascine revetment.* A *fascine* (Fig. 22) is a bundle of twigs closely bound up. There are two sizes of fascines; one size is nine inches in diameter, and about ten feet long; the other, which is generally termed a *soucisson*, is twelve inches in diameter and twenty feet long; it is chiefly used for the revetments of batteries.

76. To make a fascine, straight twigs are selected, between the thickness of the little finger and thumb, the longer the better; they should be stripped of the smaller twigs. A machine, termed a *fascine horse*, is put up, by driving two stout poles obliquely into the ground about two feet, so as to cross each other about two feet above the ground, where they are firmly tied together; as many of these supports as may be required are put up in a straight line, about eighteen inches apart; this forms the horse on which the twigs are laid to be bound together.

77. Another machine, termed a *fascine choker*,

is formed of two stout levers about five feet long, connected near their extremities by a chain or strong cord, which should be long enough to pass once round the fascine, and be drawn tight by means of the levers.

78. The twigs are laid on the horse, with their large and small ends alternating; the choker is applied to bring them together; and they are bound by *withes*, or *gads*, made of tough twigs, properly prepared by untwisting the fibres over a blaze, so as to render them pliant; or else stout rope yarn may be substituted for them. The gads are placed twelve inches apart, and every third or fourth one should be made with an end about three or four feet long, having a loop at the extremity to receive a picket through it; this picket is termed an *anchoring picket*, its object being to secure the fascine firmly to the parapet.

79. To form the revetment, the first row of fascines (Fig. 23) is imbedded about half its thickness below the tread of the banquette, and is secured by means of the anchoring pickets, and also by several pickets driven through the fas-

cine itself about twelve inches into the earth. The knots of the withes are laid inside, and the earth of the parapet is well packed behind the fascine. A second row is laid on the first, so as to make the requisite interior slope; it should break joints with the first row, and be connected with it by several pickets driven through them both. The other rows are laid with similar precautions; and the parapet is usually finished at top by a layer of sods.

80. *Hurdle revetment.* This revetment is made by driving poles, (Fig. 24,) making the same inclination as the interior slope, about eighteen inches below the tread of the banquette, and forming a wicker work, by interlacing twigs between them in a similar manner to basket work.

The poles should be nine inches apart, their diameter about one-and-a-half inches. They should be secured to the parapet by long withes and anchoring pickets. The top twigs should be bound together by withes.

81. *Gabion revetment.* The *gabion* (Fig. 25) is a round basket of a cylindrical form, without a

bottom, its height is usually two feet nine inches, and breadth two feet.

82. To form a gabion, a *directing circle* is made of two hoops, the difference between their radii being such, that, when placed concentrically, there shall be about one-and-three-quarter inches between them. They are kept in this position by placing small chips of wood between them, to which they are tied with pack thread. The directing circle is laid on the ground, and seven or nine pickets, about one inch in diameter and three feet long, are driven into the ground between the hoops, at equal distances apart; the directing circle is then slipped up midway from the bottom, and confined in that position. Twigs half an inch in diameter, and as long as they can be procured, are wattled between the pickets, like ordinary basket work; when finished within about one-and-a-half inches of the top, the gabion is placed with the other end up, the directing circle is taken off, and the gabion is completed within the same distance from the extremities of the pickets. The wicker work at the two ends is

secured by several withes, and the ends of the pickets are brought to a point when the gabion is ready for use.

83. The gabion revetment is seldom used except for the trenches in the attack of permanent works, where it is desirable to place the troops speedily under cover from the enemy's case shot and musketry. When used for field works, (Fig. 26,) a fascine is first laid slightly imbedded below the tread of the banquette; the gabion, which is placed on end, rests on this, so as to give it the requisite slope; it is filled with earth, and the parapet is raised behind it, and another fascine is laid on top, and in some cases two.

84. *Plank revetment.* This revetment may be made by driving pieces of four-inch scantling about three feet apart, two feet below the tread of the banquette, giving them the same slope as the interior slope. Behind these pieces, boards are nailed to sustain the earth.

85. *Sand bags* are sometimes used for revetments when other materials cannot be procured; though their use, in most cases, is to form a

speedy cover for a body of men. They are usually made of coarse canvass; the bag, when empty, is two feet eight inches long, and one foot two inches wide; they are three-fourths filled with earth, and the top is loosely tied. From their perishable nature they are only used for the occasion, as when troops are disembarked on an enemy's coast.

86. *Scarp revetment.* This revetment (Fig. 27) is formed of a frame work of heavy timber, and is used only for important field forts. A piece, termed a *cap*, or *cap-sill*, is imbedded in the ground along the line of the berm; other pieces, termed *land-ties*, are placed perpendicular to the cap, with which they are connected by a dove-tail joint; they are about eight or ten feet asunder. Cross pieces are halved into the land-ties about two feet from their extremities, and two square piles, about five feet long, are driven in the angles between the land-ties and cross pieces; inclined pieces, which serve as supports to the cap, are mortised into its under side at the same points as the land-ties. These supports usual-

ly receive a slope of ten perpendicular to one base; they generally rest on a *ground-sill* at the bottom of the ditch, to which they are mortised; this sill being held firm by square piles. The ground-sill might be omitted, by continuing the supports below the bottom of the ditch.

Behind this frame work, thick plank, or heavy scantling, is placed side by side, having the same slope as the supports; or else a rabate may be made in the cap and ground-sills, and the scantling be let in between the two pieces serving as a support to the cap. This is the more difficult construction, but it is the better, since, should the heavy supports be cut away, the cap will still be retained in its place.

87. Scarp revetments are sometimes formed by laying heavy timber in a horizontal position; but this method is bad, as it enables the enemy to form a foot-hold by thrusting a bayonet between the joints.

88. The length of the land-ties should be at least equal to two-thirds the depth of the ditch.

89. The counterscarp is seldom reveted. A

frame work similar to that for the scarp might be used, and thick boards, laid horizontally, be substituted for the inclined scantling.

90. When a scarp revetment is made, the excavation of the ditch must be conducted in a different manner from that already explained. In this case, after the cap-sill and land-ties are laid, the excavation is carried on to the bottom of the ditch, by removing only earth enough to allow the frame work to be put up. A scaffolding of plank is then raised in the ditch on which the earth, that remains to be excavated, is thrown, and from there on the berm.

CHAPTER VII.

ACCESSARY, OR SECONDARY MEANS OF DEFENCE.

100. THE means here adverted to usually consists of artificial obstacles, arranged for the purpose of detaining the enemy in a position where he will be greatly cut up by the fire of the work.

101. Any thing may be regarded as an obstacle to the enemy by which his attention is diverted from the assailed to his own situation; but no obstacle will be of service to the assailed which is not within striking distance of his weapons. The proper disposition, therefore, of obstacles, is in front of the ditch within short musket range.

102. Marshes, water courses, wet ditches, precipices, &c., should in this case be regarded as natural obstacles, if they are sufficient in themselves to stop the enemy's progress. But, however strong, they are not solely to be relied on.

The strongest natural position may be carried if not vigilantly guarded.

103. In placing the ground around a work in a defensive attitude, every means should be taken to reduce to the smallest possible number the points by which the enemy might approach; so that by accumulating the troops on the weak points, a more vigorous defence may be made. In making this arrangement, equal care should be given to every thing that, affording a shelter to the enemy, would enable him to approach the work unexposed to its fires. To prevent this, all hollow-roads, or dry ditches, which are not enfiladed by the principal works, should either be filled up, or else be guarded by a detachment, covered by an advanced work. All trees, underwood, hedges, enclosures, and houses, should be cut down and levelled within cannon range, and no stumps be allowed higher than two feet. Trees beyond cannon range should not be felled; or if felled, they should be burnt, to prevent the enemy's movements being concealed.

104. If there are approaches, such as perma-

nent bridges, fords, and roads, which are equally serviceable to the assailed and to the enemy, they should be guarded with peculiar care; and be exposed to the enfilading fire of a work especially erected for their defence.

105. The principal artificial obstacles are *trous-de-loup*, or *military pits; abattis; palisades; fraises; stoccades; chevaux-de-frise; small pickets; entanglements; crows-feet*; *inundations;* and *mines.*

106. *Trous-de-loup.* (Fig. 28.) These are pits in the form of an inverted truncated cone, or quadrilateral pyramid; their diameter at top is six feet, their depth six feet, and width at bottom eighteen inches. A stake is, in some cases, planted firmly in the bottom, its top being sharpened, and the point a few inches below the upper circle.

107. Trous-de-loup are generally placed in three rows, in quincunx order, a few yards in front of the ditch. They are readily laid out by means of an equilateral triangle, formed of cords, the sides of the triangle being eighteen feet; the

angular points mark the centres of the pits. The earth taken from them is spread over the ground between them, and is formed into hillocks to render the passage between as difficult as possible. If brush wood, or light hurdles, can be procured, the pits may be made narrower, and covered with the hurdles, over which a layer of earth is spread.

Trous-de-loup are sometimes placed in the ditch; in this case their upper circles touch.

108. This obstacle is principally serviceable against cavalry.

109. *Abattis.* (Fig. 21.) The large limbs of trees are selected for an abattis. The smaller branches are chopped off, and the ends, pointed and interlaced with some care, are presented towards the enemy. The large end of the limb is secured to the ground by a *crotchet-picket*, and may be partly imbedded to prevent its being readily torn up.

110. Abattis are placed in front of the ditch; in this position they must be covered from the ene-

my's fire by a small glacis. They are sometimes placed in the ditch against the counterscarp.

111. This is an excellent obstacle in a wooded country, and admits of a good defence, if a slight parapet is thrown up behind it. The parapet may be made of the trunks of trees laid on each other with a shallow ditch, or trench, behind them; the earth from which is thrown against the trunks. In an open position it may be relied on as a security against a surprise, particularly of cavalry.

112. *Palisades.* (Fig. 23.) A palisade is a stake about ten feet long and of a triangular form, each side of the triangle being eight inches. The trunks of straight trees should be selected to get out palisades. The diameter of the trunk should be from sixteen to twenty inches. The trunk is sawed into lengths of ten-and-a-half feet, and is split up into rails, each length furnishing from five to seven rails. The palisade is pointed at top, the other extremity may be charred if the wood is seasoned, otherwise the charring will be of no service. *A palisading* is a

row of palisades set in the ground, either vertically, or slightly inclined towards the enemy. To plant the palisades, a trench is dug three feet deep; they are then placed about three inches asunder, with an edge towards the enemy. Each palisade is nailed to a strip of thick plank, termed a *riband*, placed horizontally about one foot below the ground; another riband is placed eighteen inches below the top. The earth is firmly packed in the trench.

113. A palisading is sometimes used as a primary means of defence, particularly for low works. A banquette is thrown up for this purpose against it; the tread of the banquette being six feet below the top of the palisading, and four feet three inches below the upper riband.

114. As an obstacle, it is best placed at the foot of the counterscarp; the points being twelve inches below its crest, or else covered by a small glacis. In this position the palisading fulfils all the conditions of an efficient obstacle; it is under the fire of the work; covered from the enemy's

fire; will not afford a shelter to the enemy; and cannot be cut down without great difficulty.

115. *Fraise.* (Fig. 24.) This obstacle is formed of palisades, placed, in juxtaposition, either horizontally, or slightly inclined. The best position for a fraise is on the berm, or a little below it, so as to be covered by the counterscarp crest. The part of the fraise under the parapet is termed the *tail*, and is about five feet long. To make a fraise, a horizontal piece of four-inch scantling, termed a *cushion*, is first laid parallel to the berm; each palisade is nailed to this, and a thick riband is nailed on top of the fraise near the end.

116. The point of the fraise should be at least seven feet above the bottom of the ditch; and should not project beyond the foot of the scarp, so as not to shelter the enemy from logs, stones, &c., rolled from the parapet into the ditch.

117. *Stoccade.* Trunks of small trees from nine to twelve inches in diameter, and twelve feet long, are selected to form a stoccade. They are planted in juxtaposition, in a similar manner

EXPLANATIONS OF PLATE III.

Fig. Shows the profile of a work in process of construction.

Fig. 20. Shows the plan and cross section of sods.

Fig. 21. Shows the profile of a work with a sod revetment; and a plan of the sod work, A being the headers of the bottom layer, and B those of the next layer.

Fig. 22. Shows the elevation of a fascine horse A, B; and the choker C, D, applied to the fascine; F shows the dimensions of the fascine, and its anchoring gads.

Fig. 23. Shows a profile of a work with a fascine revetment, and palisades in its ditch.

Fig. 24. Shows a profile of a work with a hurdle revetment and a fraise.

Fig. 25. A shows the elevation of a gabion half finished; and B the plan of the directing circle.

Fig. 26. Shows the profile of a work with a gabion revetment; small pickets in its ditch; and an abattis in front of the ditch, covered by a glacis.

Fig. 27. Shows a plan and section of a timber revetment B B′, is the cap; D, D′, the sill; C C′, uprights let into B and D; A, A′, the land-ties, dove-tailed into B, and confined by the cross-pieces E, E′, halved into them, and by the square short piles F; G G′ are the piles to retain the sill.

Plate 3.

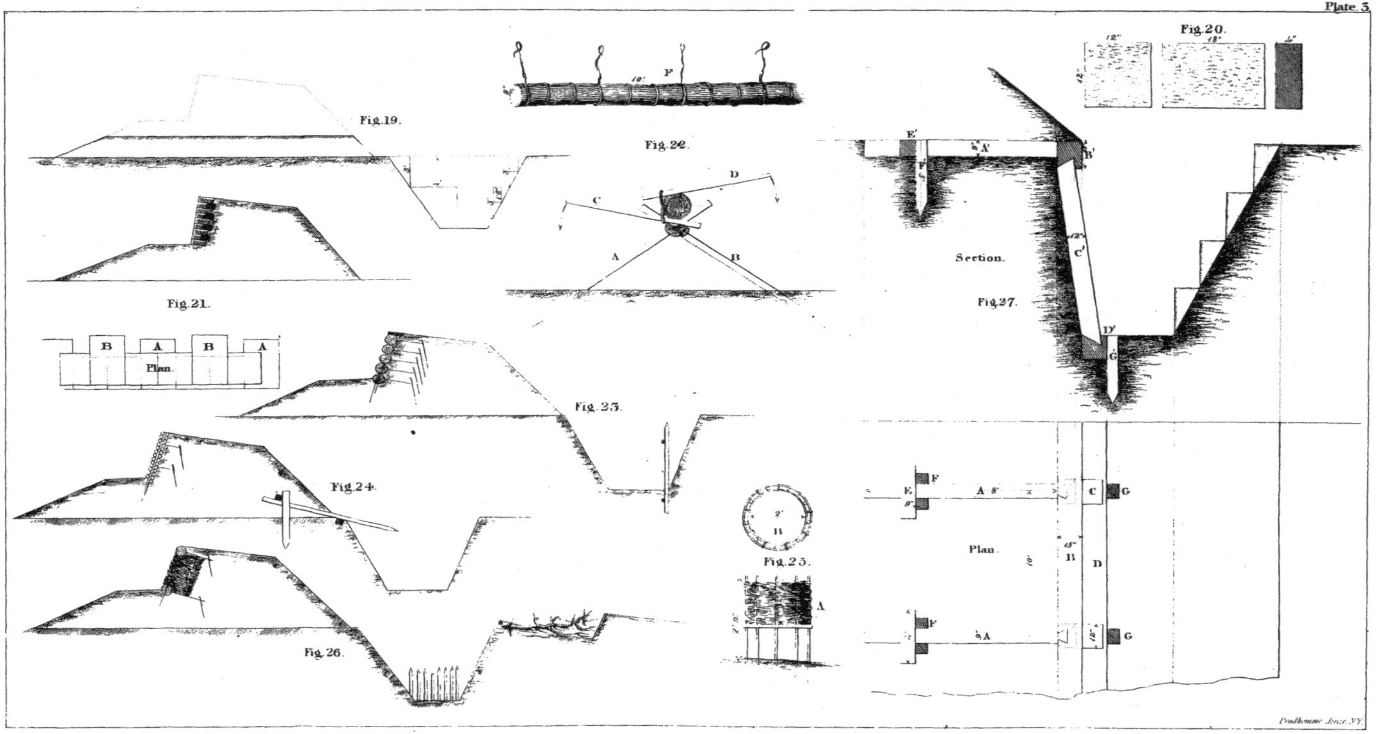
Fig. 19.
Fig. 20.
Fig. 21.
Plan.
Fig. 22.
Fig. 23.
Fig. 24.
Fig. 25.
Fig. 26.
Section.
Fig. 27.
Plan.

to a palisading, and are used for the same purposes. The manner of arranging a stoccade, which is also sometimes termed a *picket*, as a primary defence, will be described in another chapter.

118. *Chevaux-de-frise.* A cheval-de-frise (Fig. 29) consists of a horizontal piece of scantling of a square, or hexagonal form, termed the *body*, about nine feet long, which is perforated by holes two inches in diameter, and five inches apart; round staffs, ten feet long, and two inches in diameter, termed *lances*, shod with iron points, are inserted into the body, so as to project equally from it. At one end of the body a ring and chain are attached; at the other, a hook and chain; for the purpose of attaching several together, forming a chevaux-de-frise.

119. The square is the best form for the body; it requires only five-inch scantling, whereas the hexagon will require twelve-inch timber.

120. The chevaux-de-frise is not much in use as an obstacle, owing to the difficulty of making it. It is a good defence against cavalry; and on rock may supply the place of palisades; but even

here an abattis would be more effective, and more readily formed.

121. *Small pickets.* This obstacle (Fig. 26) consists of straight branches of tough wood cut into lengths of two-and-a-half or three feet. They are driven into the ground, in a quincunx order, about twelve inches apart, and project irregularly above it, not more than eighteen inches. Interlaced with cords, grape-vines, brambles, prickly shrubs, &c., they form an excellent *entanglement.*

122. The *crows-foot* is formed of four points of iron, each spike about two-and-a-half inches long, and so arranged, that when thrown on the ground one of the points will be upwards. They are a good obstacle against cavalry, but are seldom used. Boards, with sharp nails driven through them, may supply the place of crows-feet. The boards are imbedded in the ground, with the sharp points projecting a little above it.

123. *Inundations.* This obstacle is formed by damming back a shallow water-course, so as to make it overflow its valley. To be effective,

an inundation should be six feet deep. When this depth cannot be procured, trous-de-loup, or else short ditches, placed in a quincunx order, are dug, and the whole is covered with a sheet of water, which, at the ditches, must be at least six feet in depth.

124. The *dams* used to form an inundation are made of good binding earth. They cannot, in general, be raised higher than ten feet; they need not be thicker than five feet at top, unless they are exposed to a fire of artillery, in which case they should be regulated in the same way as a parapet. The slope of the dam down-stream should be the natural slope of the earth; but up-stream the slope should have a base twice the base of the natural slope. *Sluices* are made in the dams, in a similar manner to the sluices of a mill-dam, for the purpose of regulating the level of the water in the pool above, in case of heavy rains. *Waste-wiers* are also recommended, by some, for the same purpose, but it is doubtful whether they would not endanger the safety of the dam.

125. The distance of the dams apart will depend on the slope of the stream. The level of each pool should be at least eighteen inches below the top of the dam; and the depth of water below each dam should be at least six feet. These data will suffice to determine the centre line, or *axis* of each dam.

126. Artificial inundations seldom admit of being turned to an effective use, owing to the difficulties in forming them, and the ease with which they can be drained by the enemy. But when it is practicable to procure only a shallow sheet of water, it should not be neglected; because it inspires the enemy with distrust. In some cases, by damming back a brook, the water may be raised to such a level as to be conducted into the ditches of the work, so as to render some parts unassailable. The ditches, in such cases, should be made very wide, and to hold about a depth of six feet. During the winter, if the water should freeze, the ice should be broken in the middle of the ditch, and a channel of twelve feet at least, be kept open, if practicable. The ice

taken out should be piled up irregularly on each side of the channel; and, as a farther precaution against a surprise, water should be thrown on the parapet and be suffered to freeze.

127. *Mines.* Attempts at applying mines to the defence of field works have seldom proved successful, owing to the rapid character of the assault, which generally causes the mines to be sprung too soon or too late; so that the only effect that can be counted upon from their use is the panic they may inspire.

128. There is one species of mine, denominated a *stone-fougasse* which it is thought might be successfully applied to the defence of the ditches and salients of field works. To make this mine, (Fig. 30,) an inclined funnel-shaped excavation is made, to the depth of five or six feet, at the bottom of the funnel a box containing fifty-five pounds of powder is placed, with which a *powder-hose* communicates. A strong shield of wood, formed of battens well nailed together, is placed in front of the box; and three or four cubic yards of pebbles, or an equal weight of

brick bats, or other materials, are filled in against the shield. Earth is then well rammed around the shield on top and behind, to prevent the explosion from taking place in a wrong direction. A fougasse of this size, when sprung, will scatter the pebbles over a surface sixty yards in length, and seventy yards in breadth.

CHAPTER VIII.

INTERIOR ARRANGEMENTS.

129. UNDER the head of interior arrangements is comprehended all the means resorted to within the work to procure an efficient defence; to preserve the troops, and the *materiel*, from the destructive effects of the enemy's fire; and to prevent a surprise.

130. The class of constructions required for the above purposes, are *batteries; powder-magazines; traverses; shelters; enclosures for gorges*, and *outlets; interior safety-redoubt*, or *keep;* and *bridges of communication.*

131. All arrangements made for the defence, with musketry and artillery, belong to what is termed the *armament*.

132. The armament with musketry is complete, when the banquette and the interior and

superior slopes are properly arranged, to enable the soldier to deliver his fire with effect; and to mount on the parapet to meet the enemy with the bayonet. For this last purpose stout pickets may be driven into the interior slope, about midway from the bottom, and three feet apart. The armament with artillery is, in like manner, complete, when suitable means are taken to allow the guns to fire over the parapet, or through openings made in it; and when all the required accessaries are provided for the service of the guns.

133. The armament with artillery is a subject of great importance, because it cannot be applied to all classes of works. Experience has demonstrated that the most efficient employment of artillery, is in the protection it affords to the collateral salient, by a well served flank and cross fire, which shall not leave untouched a single foot of ground within its range, over which the enemy must approach. It has moreover shown, that a work with a weak profile does not afford sufficient security to warrant the use of

artillery in its defence; for artillery cannot defend itself, and such a work is too easily carried by assault to offer any hope of keeping the enemy at a distance long enough to allow the artillery to produce its effect.

134. The best position, therefore, for artillery is on the flanks and salients of a work; because from these points the salients are best protected, and the approaches best swept; and the guns should be collected at these points in batteries of several pieces; for experience has likewise shown, that it is only by opening a heavy, well-sustained fire, on the enemy's columns, that an efficient check can be given to them. If only a few files are taken off, or the shot passes over the men, it rather inspires the enemy with confidence in his safety, and with contempt for the defences.

135. *Batteries.* The term battery is usually applied to a collection of several guns; it is also used in speaking of the arrangments made of a parapet to enable the guns to fire over it, or through openings in it; as a *barbette battery*, an *embrasure battery*, &c. Two kinds of batteries are

used in the defence of intrenchments, the barbette battery and the embrasure battery.

136 The barbette is a construction, by means of which a piece can fire over a parapet. It consists of a mound of earth, thrown up against the interior slope ; the upper surface of the mound is level, or horizontal, and is two feet nine inches below the interior crest for guns of small caliber, and four feet for heavy guns. If the barbette is made along a face its length should be arranged to allow sixteen-and-a-half to eighteen feet, along the interior crest, for each gun ; and its *depth*, or the perpendicular distance from the foot of the interior slope to the rear, should be twenty-four feet, to allow room for the service of the guns. The earth of the barbette at the rear and sides receives the natural slope. To ascend the barbette, a construction, termed a *ramp*, is made ; this is an inclined plane of earth, which connects the top of the barbette with the terre-plein. The ramp is ten feet wide at top, and its slope is six base to one perpendicular. The earth at the sides receives the natural slope. The position of the

ramp may be either on one of the sides or in the rear. It should take up as little room as possible, and be convenient for the removal of the guns.

137. As the barbettes are usually placed in the salients, an arrangement is made to allow the guns to fire in the direction of the capital. The construction in this case is somewhat different from the preceding. A pan-coupé of eleven feet (Fig. 31) is first made; from the foot of the interior slope at the pan-coupé, a distance of twenty-four feet is set off along the capital; at the extremity of this line a perpendicular is drawn to the capital; and five feet are set off on this perpendicular on each side of the capital; from these points, on the perpendicular, a line is drawn perpendicular to each face respectively; the hexagonal figure, thus laid out, is the surface of the barbette for one gun. The ramp in this case is made along the capital.

138. If three or more guns are placed in the salient, a pan-coupé is formed as in the last case; and twenty-four feet (Fig. 32) are in like man-

ner set off on the capital; but instead of proceeding as in the last case, a perpendicular is drawn from this point to each face, and the pentagonal space, thus enclosed, will be taken for the gun in the salient; from the perpendiculars last set off, as many times sixteen-and-a-half feet will be set off, on the interior crest of each face, as there are guns required: this will give the length of the barbette along each face; the depth will be made twenty-four feet, and the two will be united in the salient. One or more ramps may be made as most convenient.

139. The advantages of the barbette consist in the commanding position given to the guns, and in a very wide field of fire; on these accounts the salients are the best positions for them. Their defects are, that they expose the guns and men to the enemy's artillery and sharp shooters.

Light guns, particularly howitzers, are the best for arming barbettes; because the hollow projectile of the latter is very formidable, both to the enemy's columns and to his cavalry; and when

EXPLANATIONS OF PLATE IV.

Fig. 28, 29. Shows a plan and section of an arrangement of conical trous-de-loup; small pyramidal pits, with pickets; and an advanced glacis and abattis.

Fig. 30. Shows an elevation B, and a cross section A, of a cheval-de-frise.

Fig. 31. Shows a plan and section of a stone fongasse, A A′ is the powder box, and the shield; B, B′, the stones in the funnel; C, C′, the powder hose to fire the fougasse; D, D′, the powder trough tamped with sand bags, which, with the arrangement of the earth as shown in the section, are to prevent the load from acting to the rear.

Fig. 32. Shows the plan of a barbette for one gun; A is the surface of the barbette; B that of the ramp; C the platform; D the superior slope at the pan-coupé.

NOTE.—*On the section* m n *the letters* B′, C′, D′, *correspond to the same on the plan.*

Fig. E. Shows a plat-form of beams; a is the heurter; b the wheel supports; c the trail supports; d, d cross supports of b and c, confined by pickets.

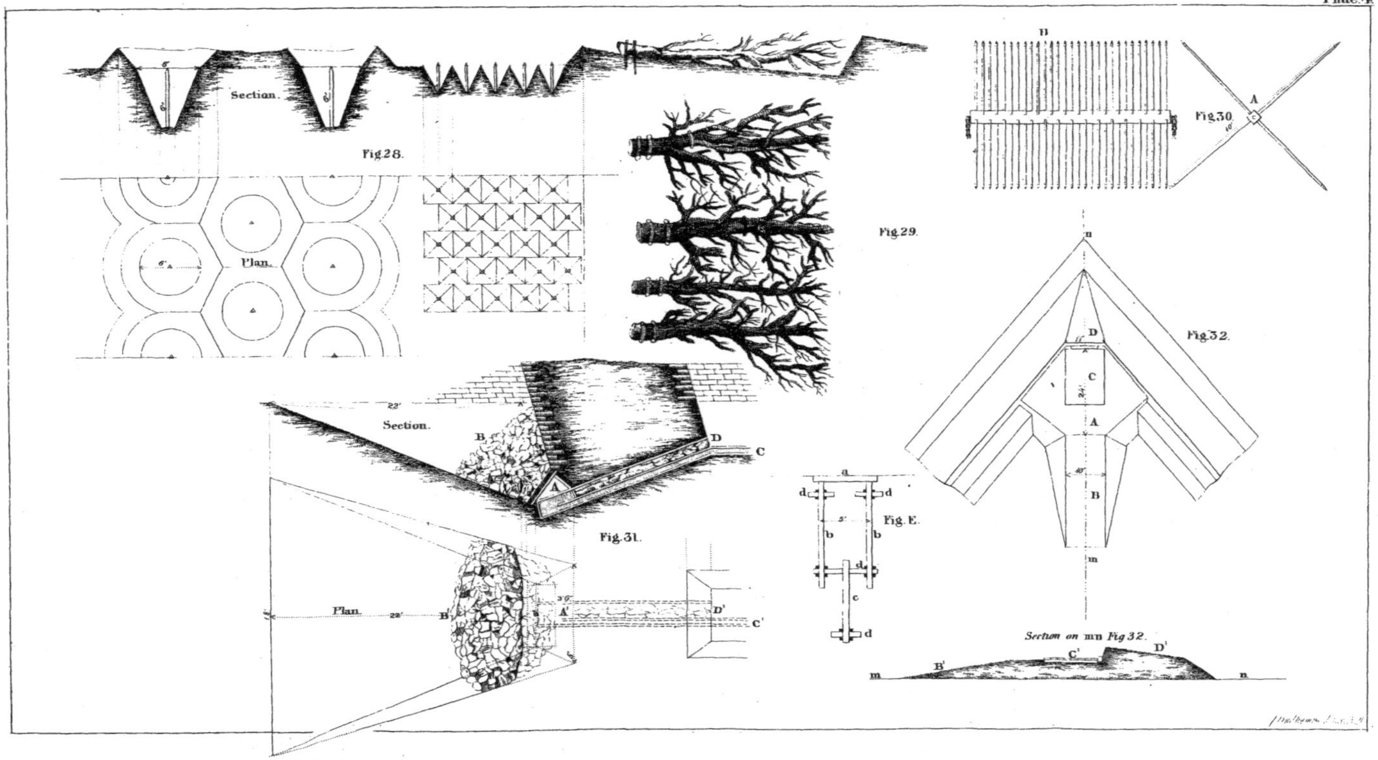
Section.
Fig. 28.
Plan.
6'
Fig. 29.
B
Fig. 30.
A
10'
n
Fig. 32.
D
C
A
B
40'
m
22'
Section.
B
D
C
A
Fig. 31.
Plan.
22'
B'
A'
D'
C'
3'0"
a
d
Fig. E.
b
c
Section on mn Fig 32.
B'
C'
D'
m
n

his batteries are opened against the salients, the light pieces can be readily withdrawn.

140. The embrasure (Fig. 33) is an opening made in the parapet for a gun to fire through. The bottom of the embrasure, termed the *sole*, is two feet nine inches, or four feet above the ground, on which the wheels of the carriage rests, according to the size of the gun; it slopes outwards to allow the gun to be fired under an inclination, the base of this slope should never be less than six times the altitude; the interior opening, termed the *mouth*, is from eighteen inches to two feet wide, according to the caliber of the gun, and is of a rectangular form; the embrasure widens towards the exterior, which widening is termed the *splay;* the manner in which the splay is regulated, is by producing the sole to the exterior slope of the parapet, and making this exterior line, measured on the sole, equal to half the distance between the inner and outer lines of the sole. This construction makes the sole a trapezoidal figure; the side of the trapezoid, on the interior being eighteen inches, or

two feet; the opposite side being equal to half the perpendicular distance between the two sides. The line which bisects the sole is termed the *directrix* of the embrasure; the sides of the embrasure, termed the *cheeks*, are laid out, by setting off two points on the exterior crest of the parapet, one on the right, the other on the left of the sole, so that the horizontal distance of these points, from the sole, shall be equal to one-third their height above it. Lines are then drawn, on the exterior slope, from these points to the extreme points of the sole; lines are in like manner drawn from the same points, on the superior slope, to the extreme points of the mouth, on the interior crest. These four lines form the boundaries of the two cheeks on the superior and exterior slopes.

141. When the directrix is perpendicular to the direction of the parapet, the embrasure is termed *direct;* when the directrix makes an acute angle with it, the embrasure is termed *oblique*.

142. The manner of laying out an oblique

embrasure is similar to the direct; the mouth is of a rectangular form, but is made wider in proportion to the obliquity, in order that the part of the embrasure, which corresponds to the muzzle of the gun, may be nearly of the same width in both the direct and oblique embrasures. The exterior width of the sole is made equal to one half the length of the directrix, measured on the sole. The cheeks are laid out as in the last case.

143. The muzzle of a gun should enter at least six inches into the embrasure, to prevent the blast from injuring the cheeks; this limits the obliquity of the directrix to 60° for long guns.

144. The height of the cheeks must not be more than four feet, for the same reason; it will therefore, in some cases, be necessary to raise the ground on which the wheels rest.

145. The parapet of a battery is, usually, termed the *epaulment.* The interior face of the epaulment, and the cheeks of the embrasures, are revetted in the usual manner. That part of the interior face which lies below the chase

of the gun is termed the *genouillere.* The mass of earth between two embrasures is termed a *merlon.*

146. The embrasures are generally cut out after the epaulment is thrown up. If their position is decided upon beforehand, they may be roughly formed at first, and be finished after the epaulment is ready.

147. The advantages of embrasures are, that the men and guns are less exposed than in a barbette battery. Their principal defects are, that they have a very limited field of fire; they weaken the parapet; and present openings through which the enemy may penetrate in an assault. Owing to their limited field of fire, they are chiefly used for the protection of particular points; as to flank a ditch, protect a salient, enfilade a road, &c. The most suitable position for them in a work is on the flanks.

148. *Platforms.* When a gun is fired often in the same direction, the ground under the wheels is soon worn into a rut; it is to prevent

this that platforms of timber are used in such cases.

149. The shape of the platform is usually a rectangle; in some cases, where a wide field of fire is to be obtained, the form is a trapezoid. The rectangular platform (Fig. 33) is ten feet wide, and seventeen feet long, for siege pieces; and nine feet wide and fifteen long, for field guns. It consists of three *sleepers* of six-inch scantling, either fifteen or seventeen feet long, which are laid perpendicular to the direction of the epaulment, and are covered with two-inch plank, twelve inches wide, and cut into lengths of nine or ten feet. Between the ends of the sleepers, and the foot of the genouillere, a piece of eight-inch scantling, nine feet long, termed a *heurter* is laid; it should project about six inches above the platform, and be bisected by the directrix. The object of the heurter is to prevent the wheels from being run against the revetment, and also to give the gun its proper direction, particularly in night firing.

150. To lay a platform, the earth on which it

is to rest should be well rammed and levelled; three trenches are then made for the sleepers, two of which should be placed under the wheels, and the middle one under the trail. The sleepers are laid flush with the ground, and firmly secured by pickets driven at their sides and ends, and the earth is solidly packed in the trench around them; the plank is then laid and secured by nails, or some other fastenings.

151. If the platform is for direct firing, with full charges, the *tail* may be made six inches higher than the front to break the recoil; in all other cases it should be horizontal.

152. A platform may be constructed simply of three pieces of timber, (Fig. 31,) one under each wheel, and one under the trail, firmly secured by pickets, and connected by cross pieces, into which they are halved.

152. For barbettes, the platform may be dispensed with; or, if used, the whole surface nearly of the barbette should be covered.

153. If the platform is made of a trapezoidal form, it will require five sleepers.

154. *Powder magazines.* The main objects to be attended to in a powder magazine are, to place it in the position least exposed to the enemy's fire; to make it shot proof; and secure the powder from moisture.

155. If there are traverses, such, for example, as are used in defiladement, the magazines may be made in them; or they may be placed at the foot of a barbette; or, in dry soils, be made partly under ground.

156. The magazine should be at least six feet high, and about the same width within; its length will depend on the quantity of ammunition. It may be constructed of fascines, gabions, or *coffer-work*, or any means found at hand may be used which will effect the end in view.

157. If fascines are used, the sides (Fig. 34) should slope outwards to resist the pressure of the earth; the fascines should be firmly secured by pickets, and anchoring withes. The top may be formed by a row of joists, of six-inch scantling, placed about two-and-a-half feet apart; these should be covered by two layers of fascines

laid side by side, and the whole be covered in by at least three feet thickness of earth. The bottom should be covered by a flooring of joists and boards; a shallow cellar being left under the flooring, with a pitch towards the door of the magazine, to allow any water that might leak through to be taken out. A thatch of straw might be used on the inside, but it is somewhat dangerous, owing to its combustibility; hides or tarpaulins are better, and will keep out the moisture more effectually.

158. A *coffer-work* is formed by making frames (Fig. 35) of six-inch scantling; each frame is composed of two uprights, termed *stanchions*, and a cap and ground-sill, well nailed together; it is six feet wide, and six feet high in the clear. These frames are placed upright, and parallel to each other, about two-and-a-half feet apart; they are covered on the top and sides by one-and-a-half-inch plank, which is termed a *sheeting*. The magazine otherwise is constructed as in the last case.

159. When gabions are used, a hole is usually dug in the ground to form a part of the maga-

zine; the gabions (Fig. 36) are placed in two rows, side by side, around the hole, and are filled with earth. The top is formed as in the case of fascines.

160. The mouth of the magizine is covered by a splinter proof shelter. This is constructed (Fig. 37) by taking scantling eight by ten inches, cut into suitable lengths, and placing it in an inclined position, so as to cover the mouth, and leave an easy access to it. The pieces, usually, are inclined 45°, and are placed side by side; they are covered by at léast two feet of earth, or sods; and hides or tarpaulins are thrown over the whole.

161. These shelters may also be used as a kind of barrack for the troops; and to store provisions, &c., in.

162. *Traverses.* Those which are constructed to cover the flanks of the guns from an enfilade fire, are usually what are termed *gabionades.* To form a gabionade, gabions are placed in a row, (Fig. 38,) side by side, enclosing a rectangular space of about twelve feet in width from

out to out, and about twenty-four feet in length, perpendicularly to the epaulment. A second row is placed within this and touching it. The whole area thus enclosed is filled in with earth, to a level with the top of the gabions. Four rows of large fascines are next laid on the gabions, to support a second tier consisting of one row. The second tier is filled in like the first, and the earth is heaped up on top, making the gabionade nearly eight feet high. It may be as well to state, that the work will be expedited by throwing up the greater part of the earth before placing the second tier.

163. *Enclosures for gorges and outlets.* A stoccade is the best enclosure for the gorge of a work. The outline, (Fig. 39,) or plan of the gorge, should be a small bastion front, for the purpose of obtaining a flank defence.

164. The trunks for the stoccade (Fig. 40) should be ten or twelve inches in diameter, and eleven feet in length. It would be best to square them on two sides, so that they may have about four inches of surface in contact. The top of

EXPLANATIONS OF PLATE V.

Fig. 33. Shows the plan of a barbette for three guns; the letters designate the same parts as in Fig. 32.

Fig. 33. (Bis.) Shows the arrangement of a Direct and Oblique Embrasure, with the common Platform.

c d E f is a Section of the Epaulment.

C I K D is the plan of the Sole; and a b is its position in elevation.

O I P K is the plan of the Mouth, and a d is its elevation.

I O G C is the plan of the left Cheek, and a d E b, its elevation.

$CD = \frac{1}{2} AB$; $GH = \frac{1}{3} EF$.

A B, and a b, show the position of the Directrix in plan and elevation.

M is the Merlon, or mass of earth between two embrasures.

Fig. 34. Shows a cross section of a Powder Magazine made of fascines and timber.

Fig. 35. A shows a longitudinal, and B a cross section of a Powder Magazine of coffer work, covered with fascines and earth.

Fig. 36. Shows a cross section of a Powder magazine, the sides of which are formed of gabions, and the roof of timber and fascines covered with earth.

Plate 5.

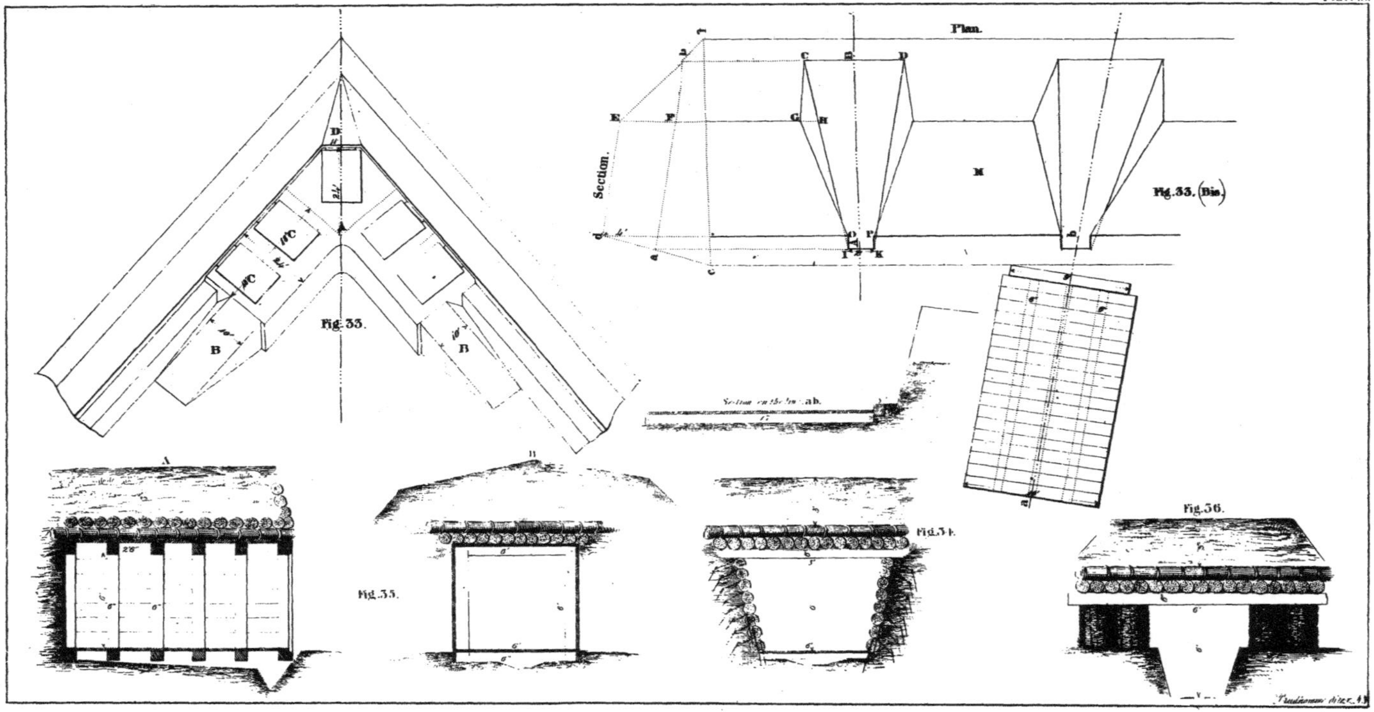
Fig. 33.
Plan.
Section.
Fig. 33. (Bis.)
Section on the line ab.
Fig. 34.
Fig. 35.
Fig. 36.

the stoccade should be at least eight feet above the ground. To arrange it for defence, a banquette should be thrown up against it on the interior; the height of the banquette one foot nine inches. A strip, about two feet in length, should be cut from the top of each trunk, with a saw, so that when two trunks are placed side by side there shall be an opening at top, between them, eight inches wide on the interior, and two-and-a-half inches on the exterior; this opening, through which the muzzle of the musket is run out, in firing, is termed a *loop-hole.* In this arrangement the bottom of the loop-holes will be six feet above the ground, on the exterior, to prevent the enemy from closing on them to stop them up, or use them in the attack.

165. About four feet in front of the stoccade, a ditch is made twelve feet wide and three feet deep. The earth from the ditch is thrown up against the stoccade, in a slope, to the level of the bottom of the loop-hole, to prevent the enemy from attempting to cut down the stoccade.

166. *Outlets* are passages made through a

parapet, or an enclosure of a gorge, for the service of the work. They should in all cases be made in the least exposed part of the work. Their width need not be more than six-and-a-half feet, when used only for the service of the work; but when they serve as a common passage for wagons, &c., in the case of the intrenchment crossing a road, they should be at least ten feet wide.

When cut through a parapet, the sides receive a slope of three perpendicular to one base, and are revetted with sods, &c.

167. A gate, termed a *barrier*, serves as an enclosure to the outlet. The frame work of the barrier is made like an ordinary gate, (Fig. 41,) consisting of two uprights, or *stiles*, a cross piece, or *rail*, at top and bottom, and a *swinging bar*, or a *diagonal brace*. Upright palisades, about seven feet long and four inches thick, are spiked to the frame about four inches apart; they are finished at top with spikes. A barrier, thus constructed, will not offer a shelter to the enemy should he attempt to cut it away. The

barrier is swung on hinges like an ordinary gate. The posts of the frame work should be very solidly braced to support the weight of the barrier.

A cheval-de-frise is sometimes used for a barrier; it presents but a trifling obstacle.

168. The outlet should be covered by a mask, thrown up either on the interior, or on the exterior, to prevent the enemy from firing through it into the work. A traverse (Fig. 42) is thrown across it, if placed on the interior. Sufficient space should be left between the traverse and parapet for the passage of a gun. The length of the traverse is arranged to prevent the enemy from firing into the work, by an oblique fire through the outlet. The traverse may be of earth or of wood; in either case it should be arranged for defence to enfilade the outlet. In some cases, and it would generally be safest, a barrier is erected between the parapet and the traverse, on each side of the outlet.

169. In very frequented passages, a redan (Figs. 48, 49, &c.) or a lunette, is thrown up on

the exterior to cover the outlet, and secure its safety in case of surprise.

170. *Safety Redoubt.* In enclosed works a place of retreat, into which the troops may retire in safety after a vigorous defence of the main work, will remove the fears of the garrison for the consequences of a successful attack of the enemy, and will inspire them with confidence to hold out to the last moment.

171. This interior work, which may very properly be termed the *keep*, can only be applied to works of large interior capacity. It may be formed of earth, or consist simply of a space enclosed by a defensive stoccade, or palisading. In either case it should be about four feet higher than the main work, to prevent the enemy from obtaining a plunging fire in it from the parapet of the main work.

172. The best arrangement for the keep is the construction termed the *block-house.* This work (Fig. 43) is made of heavy timber, either squared on two sides or four; the pieces which form the sides of the block-house are either laid horizon-

tally, and halved together at the ends, like an ordinary log-house, or else they are placed vertically, side by side, and connected at top by a cap-sill. The sides are arranged with loop-hole defences; and the top is formed by laying heavy logs, side by side, of the same thickness as those used for the sides, and covering them with earth to the depth of three feet.

173. With regard to the details of the construction, the timber for the sides should be twelve inches thick, to resist an attack of musketry, and to resist field-pieces two feet, in which case the sides are formed of two thicknesses of twelve-inch timber. If the timber is placed upright, each piece should be let into a mortise in the cap-sill; and every fourth piece of the top, at least, should be notched on the cap-sill, to prevent the sides from spreading out.

174. The plan of the block-house must conform to its object generally; it may be square or rectangular. (Fig. 44.) If flank defences are required, its plan may be that of a cross. The interior height should not be less than nine feet,

to allow ample room for loading the musket; this height will require that the timber of the sides shall be twelve feet long, in order to be firmly set in the earth. Sometimes a ground sill is placed under the uprights, but this is seldom necessary. The width may be only twelve feet in some cases, but it is better to allow twenty feet; this will admit of a camp bed of boards on each side, six-and-a-half feet wide, and a free space of seven feet. If cannon is to be used for the defence, the width must be at least twenty-four feet; this will allow eighteen feet for the service of the gun, which is generally ample, and six feet for a defence of musketry on the opposite side. A greater width than twenty-four feet cannot well be allowed, because the bearing would be too great between the sides for twelve-inch timber; and even for a width of sixteen feet it would be well to support the top pieces, by placing a *girder* under them resting on *shores.*

175. The loop-holes are three feet apart; their interior dimensions are twelve inches in height, and eight inches in width for sides twelve inches

thick; and twelve inches square for sides two feet thick. The width on the exterior, for the same thicknesses, will be two-and-a-half and four inches. The height of the loop-hole on the exterior will depend on the points to be defended; it should admit of the musket being fired under an elevation and a depression. The height of the loop-hole above the exterior ground is six feet.

176. Vents for the escape of the smoke are made over each loop-hole, between the cap-sill and the top pieces.

177. The camp bed serves also as a banquette; it is placed four feet three inches below the loop-hole, and has a slight slope of about eight inches inwards.

178. The provisions, accoutrements, &c., are placed on shelves and racks attached to the shores and girders.

179. The block-house is surrounded by a ditch, similar to the one used for a defensive stoccade. A strong door is made in one of the re-entering angles, and a slight bridge leads from it across the ditch.

180. It has been proposed to place a slight parapet of earth on top of the block-house. It is thought that this accumulation of earth would be too heavy for the timbers, independently of leaving but little space for the defence. Perhaps a better arrangement might be made on top, similar to a defensive stoccade, the uprights being secured at bottom, between two pieces resting on the top pieces, and held firm by an arrangement of riband pieces and braces.

181. It has also been proposed to place the interior and exterior rows of uprights three feet apart, and to fill in between them with closely packed earth, for a defence against artillery. This method has been tried, and was found to be less solid than the one here laid down, independently of being more difficult to construct.

182. The top pieces should in no case project more than twelve inches beyond the sides to admit of logs, &c., being rolled over on the enemy.

183. The block-house is sometimes arranged with two stories, the corners or the sides of the upper story projecting over the sides of the lower.

Either of these methods are sufficient for the defence of the lower story ; but the first is the best to procure a fire in the direction of its angles. It can only be used, however, as a defence against infantry.

184. When artillery cannot be brought to bear against the top of the block-house, it may be constructed like an ordinary floor, and be covered with nine or twelve inches of earth to guard against fire.

185. The applications of wood to the purposes of defence is one of paramount importance in our country. A block-house, surrounded by a defensive stoccade, is impregnable to the attack of infantry, if properly defended, and is therefore peculiarly suitable to either wooded or mountainous positions, where a train of artillery cannot be taken without great labor, owing to the impediments that may be thrown in its way, by rendering the roads impassable from obstructions easily obtained. In positions covered by extensive earthen works, such as those that would be required for the defence of the towns on our sea-

board, and which would be occupied during a war, a defensive arrangement of the barracks for the troops, so that they might serve, in case of the main works being forced, as rallying points, under cover of which the main body of troops may retreat with safety, is a subject that commends itself to the serious attention of the engineer. From the details already entered into, an efficient combination for this purpose will suggest itself to the reader, without entering farther into particulars.

186. *Ditch defences.* The surest defence for a ditch is a good flanking arrangement of the work itself; but as this is, in many cases, impracticable, owing either to the relief, or to the plan, flank defences must be procured by a construction made in the ditch. Two methods are resorted to for this purpose; the one, termed a *caponnière;* the other a *counterscarp gallery.*

187. The caponnière is a work made across the ditch, and may be either single or double. A single caponnière is nothing more than a defensive stoccade, or palisading, made at the extremity of

a ditch, as in the case of a redan. It will obviously be of no service, unless the enemy is forced to attack it in front.

188. A double caponnière (Fig. 45) is arranged to fire in two directions, and is usually placed at the middle of the ditch which it is to protect. It is made in all respects like a block-house with upright sides; its width may be only eight feet, and its height the same.

189. The bottom of the caponnière may be on the same level as the ditch, or below it; in the latter case, the loop holes should not be more than eighteen inches above the level of the ditch, to prevent the enemy from using them against the assailed within. In the former, a small ditch should be made round the caponnière, and the earth from the ditch be thrown against its sides.

190. The plan of the caponnière (Fig. 46) should be arranged to admit of being flanked by a defensive stoccade, placed at the foot of the scarp; for this purpose the end next to the counterscarp should be in the form of a salient flanked by the stoccade. To prevent the enemy

from jumping on the caponnière, from the crest of the counterscarp, a space of at least twelve feet should be left between the two. Moreover, the top of the caponnière should be covered from the enemy's artillery, either by the counterscarp crest, or by a glacis.

191. The entrance to the caponnière, from the work, may be a timber gallery under the parapet, formed of frames, and covered in by a sheeting, similar to the construction used for a powder magazine.

192. The counterscarp gallery (Fig. 47) consists of a frame work, covered in with a sheeting, which is placed within the counterscarp at the salients. The front of the gallery is made of nine or ten-inch scantling, placed upright, and arranged with loop-hole defences; these pieces are connected at top by a cap-sill. Cross pieces are notched on the cap-sill, about three feet apart; they are supported by shores placed four feet from the front piece. The cross pieces may project three feet beyond the shores, and, if necessary, be braced from the shores. The gallery is

EXPLANATIONS OF PLATE VI.

Fig. 37. Shows a cross section of a Splinter Proof Shelter laid against the side of a Traverse. The traverse is revetted with fascines, and the timber work is covered with sand bags and tarpaulins, or raw hides.

Fig. 38. Shows the cross section of a Gabionade.

Fig. 39. Shows the plan of a Lunette, with its gorge enclosed by a Stoccade, the plan of which is a small bastion front.

Fig. 40. Shows the section and exterior elevation of a Defensive Stoccade.

Fig. 41. Shows the plan and elevation of a Barrier.

A, A are the Posts to which the Gate is swung, and attached, when closed;

B, B are the Uprights of the Gate.

C, C′ the upper and lower Cross Pieces.

D the Diagonal Brace.

E the Bar of the Gate.

Fig. 42. Shows a plan and section of the Outlet through the face of a work, covered by an interior Traverse.

O is the Outlet or Passage through the parapet.

P the Passage between the Traverse T, and the foot of the banquette.

T is the Traverse, supposed of earth; its banquette being ascended by steps, instead of an ordinary slope, to save room.

Figs. 43, 44. Show the plan and section of a block-house of upright timber. The plan is made to exhibit a portion of the top complete; the timber covering of the top; the arrangement of the cap pieces; a plan of the loopholes; and a plan of the camp-bed. Fig. 43 exhibits, in like manner, a cross section of the block-house and ditch; with an interior and exterior elevation.

Fig. 45. Shows a cross section of a double caponière, with loopholes near the bottom of the ditch.

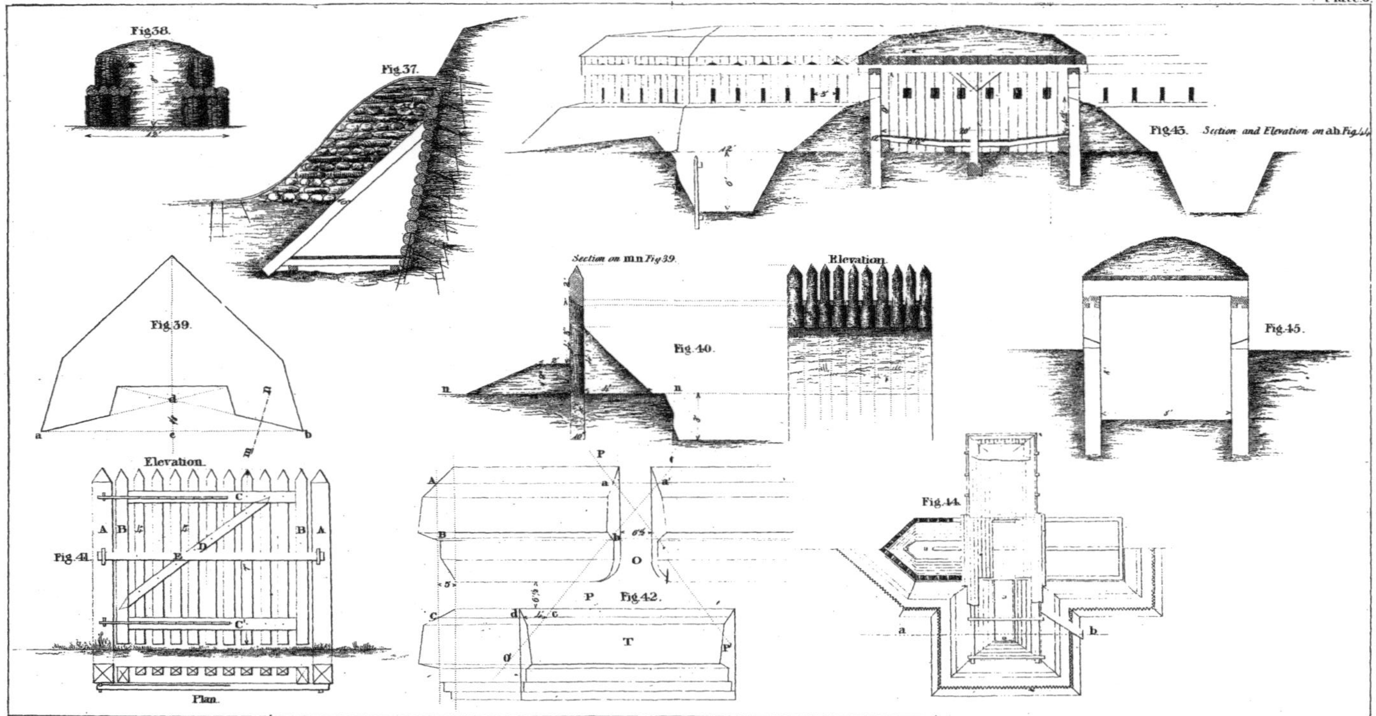
Fig. 38.
Fig. 37.
Fig. 43. Section and Elevation on ab Fig. 44.
Fig. 39.
Section on mn Fig. 39.
Fig. 40.
Elevation.
Fig. 45.
Elevation.
Fig. 41.
Plan.
Fig. 42.
Fig. 44.

covered in on top by one-and-a-half-inch sheeting; and is covered in behind, in a similar manner, only to the height of five feet above the bottom. This arrangement gives a free space behind the back sheeting for the play of the rammer in loading. The height of gallery may be only seven feet; its width, according to the foregoing arrangements, is four feet. It should be covered on top by at least three feet of earth. The level of the gallery should be the same as the ditch; and there should be a small ditch in front of it, to prevent the enemy from closing on the loop holes, or obstructing their fire by filling the ditch in front of them by means of sand bags, fascines, &c. The entrance to the gallery is by a narrow door.

CHAPTER IX.

LINES OR COMPLEX INTRENCHMENTS.

193. All dispositions made to cover extended positions, and which present a front but in one direction to the enemy, are termed *Lines*. There are two classes of lines—*Continued Lines*, and *Lines with Intervals*. Continued lines present no openings through which the enemy might penetrate except the ordinary outlets. Lines with intervals consist of detached works, which are enclosed partly, or throughout their perimeters, arranged in defensive relations with each other; and presenting wide intervals between them defended only by their fire.

194. The same general principles apply to lines as to other intrenchments; but, from their great extent, they usually receive a slight relief, and the simplest angular figures are adopted for

their plan; and in laying them out, the engineer should avail himself of all the natural obstacles, presented by the position, so as to diminish the labor of erecting artificial ones.

195. *Continued Lines.* The simplest arrangement for a continued line, consists in a series of redans (Fig. 48) connected by straight curtains; it is termed the *redan line.* The faces of the redans are sixty yards in length; their salient angles 60°; and the distance between their capitals one hundred and eighty yards.

196. This combination will place the salients at one hundred and sixty yards from the collateral re-enterings. An inspection of this system shows that the ditches are not flanked; that the salients are not well protected owing to the cross fire leaving a considerable sector without fire in front of them; that the curtains, which, from their position, are the strongest points, are the best defended, and in turn they afford no protection to the faces. All these defects become more sensible as the redans are placed farther apart.

197. To remedy the defects of the redan line,

it was proposed to break the curtains forward (Fig. 49) so as to form two branches, one perpendicular to the face of each redan. This suggestion led to the *tenaille line*, which consists of a combination of small and large redans, or simply of redans of the same size, forming salient and re-entering angles. This combination is superior to the redan line. The salients are protected by a cross fire, and the ditches of the large redans are partially flanked; moreover, it presents fewer assailable points, on a given front than the redan line; and its retired parts afford good positions for artillery.

198. The faces of the large redans should not exceed one hundred and sixty yards; and their salient angle should not be less than 60°. The faces of the small redans should not be greater than forty yards, and should be perpendicular to those of the large redans. These combinations will give two limits for the length of the capitals of the large redans, and for their distance apart. When the salient angles are 60° the length of the capitals will be about

one hundred and thirty-eight yards, and the distance between them two hundred and twenty-eight yards. When the salient angle of the small redan is 60°, the capitals of the large redans will be eighty yards long, and the distance between them three hundred and sixteen yards. In the first case there will be a greater number of assailable points on a given front, but the re-enterings will be the stronger; in the second case the reverse of this will happen. Of course the size of the redans will not be restricted to the limits here laid down. This system is defective, in presenting very long faces to an enfilade fire, and in taking up a considerable depth of ground, from the salients to the re-enterings, which restricts its application to particular localities.

199. The *indented line* (Fig. 50) is principally used in place of a straight curtain between two advanced works, which are too far apart to protect each other and the space between them.

200. When the ground between the advanced works is level, or nearly so, the branches of the

cremaillère form salient and re-entering angles, which are on the same right lines. The long branches alternate from the middle point, where either a salient or re-entering angle is formed, the latter is preferable, as it is strongest, and may be arranged with flanks and a curtain, which will be better situated for defence than the two faces forming a salient angle. When there is a valley between the two advanced works, (Fig. 51,) the branches of the cremaillère should conform to the slopes of the ground ; the long branches should be thrown back, so that their prolongation shall fall within the salients of the advanced works, to avoid an enfilading view of the enemy. The middle point may be arranged as in the preceding case.

201. The cremaillère line is more easily adapted to irregular sites, particularly to the sides of hills, or similar slopes, than either of the preceding dispositions, because it requires but a slight depth of ground.

202. Owing to the imperfect flanking arrangements of the preceding systems, it has been pro-

EXPLANATIONS OF PLATE VII.

Fig. 46. Shows a plan A of a double caponière, with flanking arrangements, the postern B, leading from the main work into it; the small ditch C, around it; and an abattis D, on the scarp, where it is joined by the caponière.

Fig. 47. Shows a cross section A, of a Counterscarp Gallery; an exterior elevation B of the gallery; and a cross section C, of a timber revetment formed of a frame work and thick plank, laid horizontally, behind the uprights.

Fig. 48. Plan of a Redan Line, with a passage A, in the curtain, covered by a small redan.

Fig. 49. Plan of a Tenaille Line.

Fig. 50. Plan of a Crémaillière Line between two salient priest-caps.

Fig. 51. Plan of a Converging Crémaillière line across a valley.

Fig. 46.

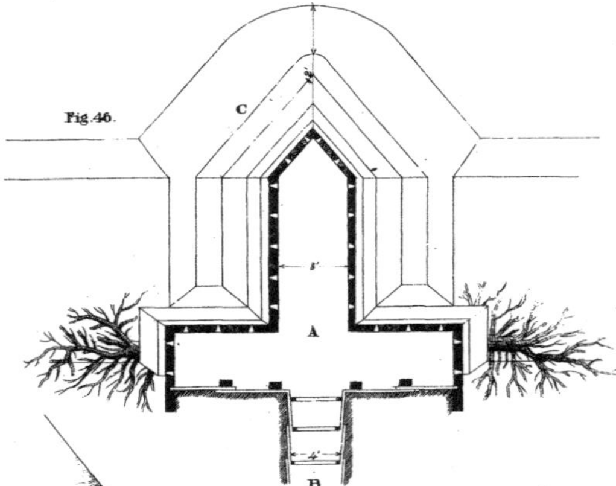

Fig. 47.

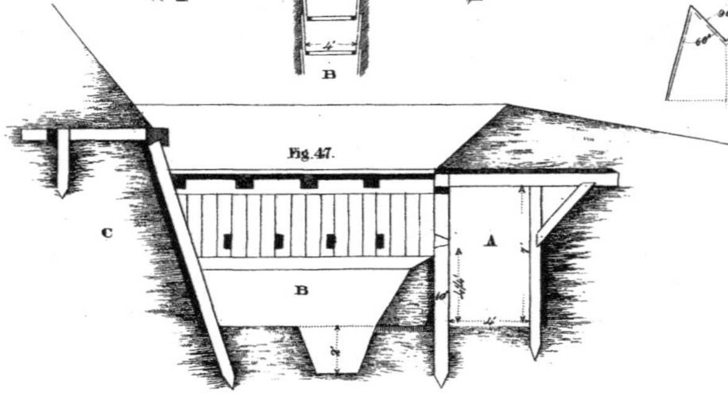

Fig. 48.

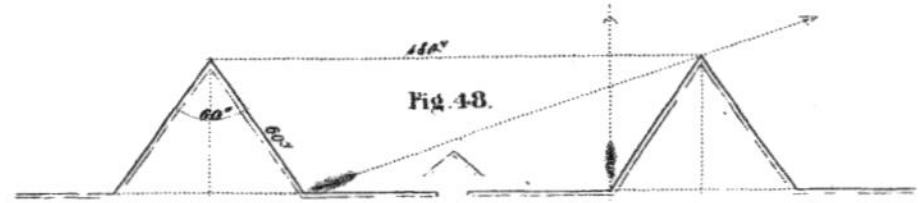

Fig. 49.

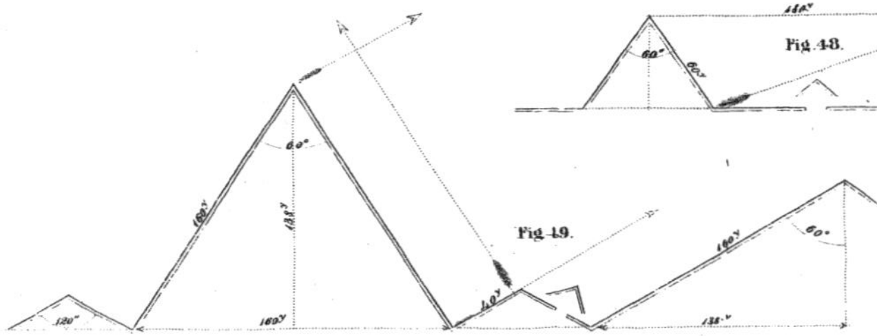

Fig. 50.

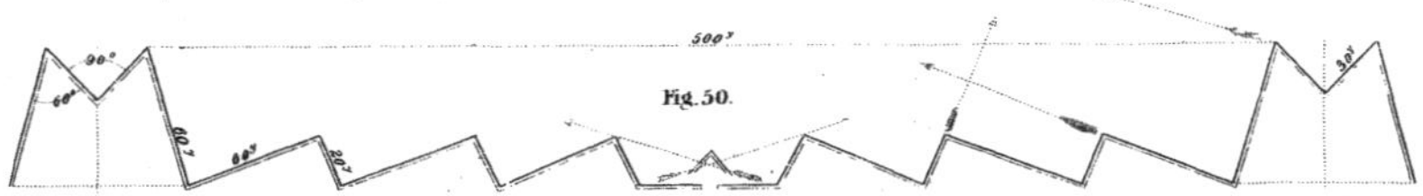

Fig. 51.

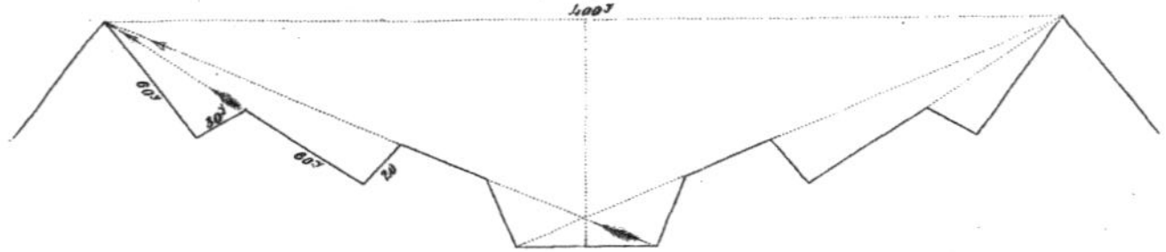

posed to use bastion lines. They are laid out by placing the salients two hundred and fifty yards apart, and making the perpendicular of the front equal to one sixth.

203. Another arrangement of the bastion line is the one termed *with double flanks.* (Fig. 52.) The salients, in this case, should be between four hundred and five hundred yards apart. The figure explains itself. By this arrangement there are fewer assailable points on the same front; one of the bastions is placed in a strong re-entering; and the salients of the advanced bastions are protected by the flank fire of the collateral advanced bastion, and also of the retired bastion.

204. The principal objection to the bastion line is its great development, and the consequent increase of labor and time for its construction.

205. Continued lines are not suited to an active defence; and this is a grave objection to their use. The enemy, if repulsed by their fire, can retreat in good order, and renew the assault at a more

opportune moment; because the assailed, if they attempt a sortie, must defile through narrow outlets, and present a feeble front to the enemy during the operation of defiling. They are, however, an admirable defence for irregular troops, owing to the confidence which they inspire. They also serve to guard against a surprise, and to prevent the predatory excursions of small detachments of the enemy.

206. *Lines with intervals.* The detatched works used for these lines may be either lunettes, or square redoubts; (Fig. 53;) their salients are two hundred and fifty yards apart, and their salient angles may usually be 90°. The faces about sixty yards, and the flanks about forty yards, are so arranged as to sweep the ground in front of the salients of the collateral works. In the rear of the first line, and opposite to the intervals, redans are placed to flank the faces of the first line; the faces of the redans may be about thirty yards. About two hundred yards in the rear of the first line, and opposite the lunettes, cavalry epaulments are thrown up, to cover

squadrons of cavalry to act on the flanks of the enemy, when his columns are shaken by the fire of the works.

207. The works of the first line are large enough to contain from three hundred to four hundred men, with a field battery in each, for their defence. The redans are good positions for batteries to flank the first line. Redoubts present an advantage over lunettes, because they are equally strong throughout, but this is accompanied by the defect that their rear faces may be used against the assailed if the enemy should carry them. The gorges of the lunettes should be enclosed by an abattis, palisading, &c., which will not afford a shelter to the enemy should he get possession of them.

208. A system of defence has been proposed by General *Rogniat*, the present *Chief* of the *French Engineer Corps*, in his work *Considérations sur l'art de la Guerre*, the spirit of the arrangement of which partakes both of the bastion line with double flanks, and of the line with intervals.

209. Points two hundred and fifty yards apart are taken for the salients of the lunettes; their faces and flanks are placed in defensive relations; and between them a redan, with a pan-coupé, is placed to flank the faces, without intercepting the fire of the flanks; a straight curtain is carried from the redan, and leaves an interval of ten yards between it and the flanks of the lunettes for sorties.

210. With regard to the profiles, the lunettes receive the minimum profile both for the parapet and ditch. The redans are simple epaulments to cover cannon fired in barbette; and the curtains consist of a trench with the earth thrown in front to form a parapet, which is so arranged that the infantry may march from the trench in order of battle over it.

211. The advantages claimed for this system are, first the short time required to form the works, by which an army may entrench its field of battle in one night; second, the lunettes form the first line of the order of battle, and contain only infantry, and the batteries are placed in the

redans, where they are most secure, protect the lunettes, and withdraw the fire of the enemy's artillery from the lunettes ; third, the curtains are defended by infantry, who can sally from them at a moment's warning and, aided by the light artillery and cavalry, who débouche through the intervals between the curtains and lunettes, attack the enemy in flank. If the flanks of his position are not secured, by natural obstacles, he proposes to throw up a strong square redoubt on each flank, and to place a heavy battery in the interval between the redoubt and the adjacent lunette.

212. Lines with intervals are peculiarly adapted to well disciplined and active troops. The works thrown in advance constitute the first line of the order of battle, against which the first shock of the enemy is partially thrown away, and he dare not attempt to neglect them, for an endeavour to penetrate through the intervals would expose his flanks to a close and deadly cross fire. If the enemy is repulsed, the main body of the army, which is drawn up in rear of

the works, immediately assumes the offensive, and, by a vigorous advance movement, charges the enemy in turn, relying on the works to cover its retreat if driven back.

213. In every combination of this nature the flanks are the weak points; they should rest, if practicable, on some unassailable point, as a marsh, river, &c.; otherwise very strong works should be thrown up for their protection.

214. It has been proposed, by some writers, to throw up several lines of detached works for the defence of a position; so that the troops in the first line may retreat under cover of the second, and so on. This arrangement, in the first place, can seldom be made, without *weakening the order of battle*, and therefore weakening the defence, by too great a dissemination of the troops. Moreover, in works of great extent there never can be that concert, which is so essential to a vigorous defence, from the impracticability to direct it properly. The troops, destined to act offensively against the enemy if repulsed, are too far in the rear to be brought up in time;

EXPLANATIONS OF PLATE VIII.

Fig. 52 Plan of a Bastion Line with double flanks.

Fig. 53. Plan of a Line with Intervals of lunettes, or square redoubts, flanked by retired redans; with cavalry epaulments in rear.

Fig. 54. Shows the plan, and profiles, of *General* ROGNIAT'S Line.

Plate 8.

Fig. 52.

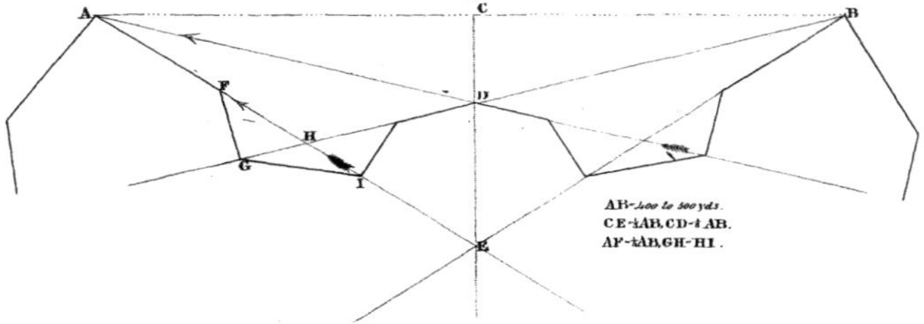

Fig. 53.

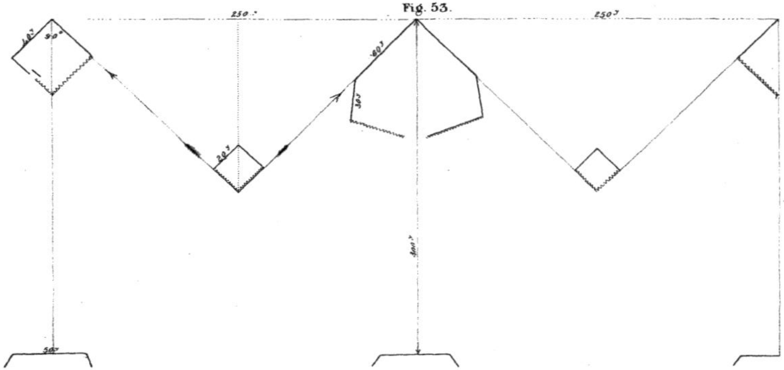

Fig. 54.

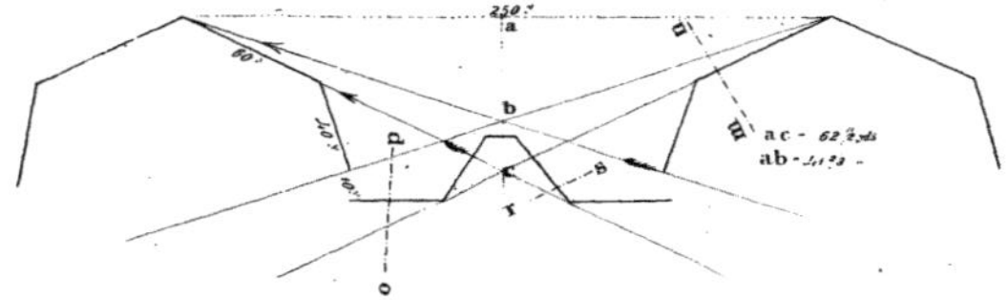

Section on mn. Fig. 54.

Section on op. Fig. 54.

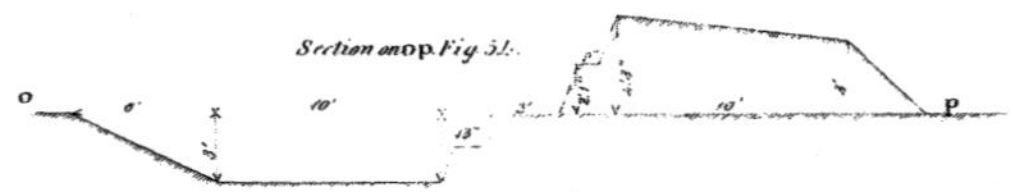

Section on rs. Fig. 54.

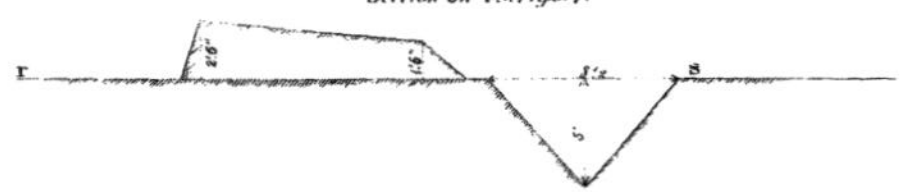

and the ground being greatly cut up, by such a multiplication of works, will render the manœuvres slow and difficult. Besides, *a very capital objection in war*, the time and labor required to throw up so many works are altogether beyond what can be disposed of in the ordinary circumstances of an army.

CHAPTER X.

THE ADAPTATION OF INTRENCHMENTS TO PARTICULAR LOCALITIES.

215. THE object of this chapter is to lay down some general rules respecting the manner of combining artificial and natural obstacles, so as to draw the greatest resources from both in strengthening a position.

216. *Mountainous sites.* The crests, and gorges, are the most important military features of a mountainous position. It is through the latter that the roads are made, and the former, from their elevation, command the latter. The crests should therefore never be abandoned to the enemy, although from their position, or distance, they may not directly overlook the gorges; for, independently of the real advantage of position, which the enemy would thus acquire, he would

possess a relative advantage in the moral effect produced on troops when they find themselves in a commanded position.

217. If the base of the mountain does not stretch out too far from the summit to admit of a sure retreat on the latter, works may be thrown up for the defence of the base, with intermediate works between the base and the summit placed on the secondary ridges, or other commanding points. But if the distance between the summit and base is great, and particularly, if it is decided beforehand to retreat upon the summits, in case of disaster, then the base should be disregarded.

218. The works thrown up for the defence of the summit should be laid out on the brow of the height, for the purpose of overlooking and guarding its sides. In planning the works, the re-entering parts of the brow should be arranged to obtain a cross and flank fire on the slopes in front of the salient points; and particular regard should be had to the inclination of the slopes, in arranging the relief and the means of defence.

219. Very steep slopes will not admit of a defence with artillery, because the gun cannot be fired under a greater depression than one-sixth, and unless the shot take effect the enemy will be inspirited to advance, confiding in the safety of his position. In slopes of this character the works may consist simply of a parapet, (Fig. 55,) in the form of a glacis, without any ditch, the earth for the parapet being taken from an interior trench; in some cases a dry stone wall may be substituted for an earthen parapet. An abattis may be formed in front of the parapet within close musket range; and heavy round logs, or large masses of rock, be arranged along the parapet, ready to be rolled over on the enemy should he break through the abattis. Steep escarpments of rock are generally considered inaccessible; but those points should never be left to their own strength. It is always prudent to post a small detachment to frustrate an attempt of the enemy to surprise them.

220. A steep natural slope, (Figs. 56, 57,) may

be readily made inaccessible by cutting away the face of the eminence.

221. It may, in some cases, be indispensably necessary to guard certain points at the base of a mountain, as, for example, where the base is washed by a river, over which there is an important ferry. Under such circumstances the point to be guarded should be protected by a strong work; and a chain of posts, placed on the most commanding points between the summit and the base, should connect the two. These posts should, when practicable, be placed in defensive relations, and in all cases their fire should sweep all the ground between the two principal points. The interior of the posts most advanced, should be exposed to the fire of those in their rear, in order that the enemy may be driven out, should he succeed in forcing his way into any one. As these posts will require a considerable detachment for their defence, care should be taken not to multiply their number unnecessarily, and never at the expense of the main defence.

222. All communications, leading through the

mountains, should be carefully guarded, both at their outlets and at the most suitable intermediate points for defence; otherwise the most respectable positions will be liable to be turned by the enemy. If the communications are not of use to the assailed, they may be barred by a line of abattis, or by an artificial inundation, &c.; and they should be watched by a detachment of light troops, whose retreat on the main works should be secured in case of an attack by superior forces.

223. If the communications are of use to the assailed they should be defended by intrenchments, which should command and enfilade them in the most effectual manner.

224. *Forests.* An abattis is the natural intrenchment of forests. The position of the abattis will depend on circumstances. If light troops are in observation in advance of the forest, they may be covered by a line of abattis, to prevent a surprise, and to enable the corps, if driven back, to retire under cover of the forest. Other lines of abattis may be formed at suitable points

in the forest itself. If the forest is in front of intrenchments, the trees must be felled within cannon range, and an abattis be formed.

225. *Marshes.* An impassable marsh, or bog, may serve to render a weak point unassailable, if the enemy can neither flank nor turn it. Like all points secured by natural obstacles, it should be vigilantly guarded to frustrate any attempt to carry it by surprise.

226. *Rivers.* The works planned for the defence of a river will depend on the object to be attained, whether it be simply to prevent the enemy from crossing, or to give the assailed the means of a secure communication with the opposite shore.

227. The points most favorable to the passage of the enemy are fords; and when the river is not fordable, the points where an elbow is formed, the re-entering being towards the enemy. To guard these weak points, works should be placed in a suitable position to prevent the enemy from approaching the opposite bank; and in the case of a ford, the plan of the works (Fig. 58)

should be so arranged that their fire can be concentrated on the ford; and, if the assailed have cavalry, a free space should be left between the water and the works for the cavalry to act on; the object being to charge the enemy's column before the troops have time to form after crossing the ford.

228. To keep open a free communication with the opposite shore, it will be necessary to throw up works there, of sufficient strength to allow the assailed time to effect a safe retreat, should they be attacked by superior forces. As these works serve to cover the bridges in their rear they are termed *têtes-de-pont*, or *bridge-heads*. The best points to erect a bridge-head are the bends, or elbows of the river, the re-entering being towards the assailed. The reasons for selecting these points are, that the bridge-head may be protected by a good flank and cross fire from the opposite shore, which from its shape is most favorable for this purpose; secondly, from the manner in which elbows are formed, the point occupied by the bridge-head will commonly

be commanded by the opposite shore, and should the enemy succeed in obtaining possession of the bridge-head, he will have all the disadvantages of a commanded position; thirdly, the elbow is not only unfavorable to the enemy, by preventing him from placing a battery in a position from which the bridge might be destroyed by artillery, but it also prevents floating bodies, thrown into the current by the enemy with a view to destroy the bridge, either by their shock, or by fire, from coming directly in contact with it; as the chances are, that those bodies will strike one or the other shore before getting into the bend.

229. The plan of a bridge head should be carefully studied on the ground; and it should, as far as practicable, satisfy the following conditions.

1st. It should admit of a defence until all the troops have effected a safe passage.

2d. It should cover the bridge from the enemy's artillery; so that the retreat may not be cut off by the destruction of the bridge.

3d. It should be suited to the end in view; if, for example, its object is to afford the means to small detachments of making incursions on the opposite shore, a small unimportant work will be all that is required; but if a large corps is to be passed over, either in retreat, or to act offensively, then the works should be arranged with wide intervals, to allow the troops to débouche in mass, and display readily in order of battle.

4th. The flanks of the works should rest upon the banks, to prevent their being turned; and, when practicable, they should be protected by flanking arrangements from the opposite shore.

5th. A strong interior redoubt should secure the bridge against any attempt on the part of the enemy to obtain possession of it by storming the works.

230. The practicability of obtaining flanking arrangements from the opposite shore will materially affect the plan of the bridge-head. If the breadth of the river is over one hundred yards, not much reliance can be placed on a flanking arrangement for musketry; and if the breadth is from six hundred to eight hundred yards, it will admit of a very effective flank fire of artillery, but

not greatly beyond this. For an unimportant work, therefore, which can be flanked by musketry from the opposite shore, a redan or a lunette, will be a very suitable form; and if the work can only be flanked by cannon, a priest-cap, (Fig. 59,) or a redan, with *crotchets*, or flanks, near the extremities of the faces, and perpendicular to them, arranged for musketry, will be a very suitable form. Neither of these works, however, admit of sufficient strength to cover a very important point, whose loss might compromise the safety of an army, or the success of a campaign. The best arrangement for this purpose is either a *simple* or a *complex crown.*

231. The simple crown (Fig. 60) consists of a central bastion, and two half bastions near the banks. In planning this work, the flanks which protect the central bastion should be longer than the other two, as it is hardly probable that the enemy will attempt an attack on any other point but the salient. The batteries on the opposite shore should sweep the ground immediately in front of the faces, and cross their fire in front of the salient.

232. The complex crown-work consists of a polygon of three or more sides, on each of which a bastion front is constructed.

233. If the bridge-head is to cover the manœuvres of a large army, either advancing, or retreating, a strong simple crown, (Fig. 60,) with a system of detached lunettes, about six hundred yards in front of it, present a very suitable arrangement. The lunettes may be arranged as in the system of General *Rogniat*. The central bastion of the crown-work should be armed with a strong battery of heavy guns to protect the lunettes; and heavy batteries, on the opposite shore, should sweep the ground between the lunettes and crown-work.

234. If there are islands in the river near the works, they may be fortified with advantage to flank them.

235. Besides the arrangements already mentioned intervals of from ten to twenty yards should be left between the shore and the works, for the troops to defile through; the interior of the work is covered by a traverse in rear of the

interval. A small defensive stoccade should be formed immediately at the head of each bridge, to enable a company of picked men to defend it until the bridge can be cast loose from the shore.

236. An interval of at least one hundred yards should be left between the bridges, if more than one is used, and about the same distance should be left between each bridge and the wings of the work.

237. *Villages.* A village may play a very important part in an action, if suitable arrangements are made for placing it in a defensive attitude, to prevent the enemy from carrying it by storm. If the village is not too far in front of the line of battle, it should be strongly occupied; its rear being left open to receive succours. But if, by occupying the village, a very salient, and therefore a very weak point, would be presented to the enemy, then it would be best to demolish it; to prevent the enemy from taking advantage of it, to cover his manœuvres whilst advancing, or from throwing a detachment into it, to cover his retreat if repulsed.

238. To admit of a good defence, the houses of the village should be built of brick or stone, and not be too scattered. The enclosures of the gardens should consist of brick, stone, or mud walls, or a thick-set hedge. Wooden villages do not admit of a good defence, on account of the ease with which they may be burnt; but they should not on this account be abandoned to the enemy; for they offer a shelter, if only a temporary one, to troops, and by checking the enemy only for a short time, may render important service. Villages consisting of log-houses are the most defensible.

239. *Military Posts.* The term *military post* is applied to isolated positions, occupied by small detachments, for the purpose of guarding particular points which are of importance during the operations of a campaign, or for a longer or shorter period. These positions are frequently villages, farm-houses, &c.

240. The officer charged with placing a village in a defensive attitude, should first proceed to a careful examination of its environs, for the pur-

pose of ascertaining what natural obstacles, and what facilities, they present to the approach of the enemy. The next points to be examined, are the walls, hedges, &c., of enclosures, which may be turned to a useful account for the defence, or which might serve as a shelter to the enemy. After having finished this examination, he will next proceed to lay out his works; arranging their plan so as to draw every possible advantage from the natural and artificial obstacles at hand, to render certain points inaccessible, and to procure a shelter for his troops, and flanking arrangements by means of the walls, hedges, &c. If there should be danger of an attack before these works can be completed, the roads leading to the village, by which the enemy might approach, should be broken up; and cannon should be placed in the best position to guard the most accessible points. The streets of the village should be barricaded, and the houses and walls, in the vicinity of the barricades, should be placed in a defensive attitude. In taking these preparatory measures against a sudden attack, any means that will afford the troops a cover from the enemy's fire should be resorted to; bales of cotton, or wool, casks set side by side, and filled with earth, piles of timber, &c., have been used with great success under such circumstances.

241. The works that surround the village should be placed so far from the houses that the troops shall not be incommoded either by the splinters occasioned by the enemy's artillery, or by the flames and smoke, should the houses be set on fire.

242. *Hedges.* A thin-set hedge cannot be placed in a good state of defence, and should therefore be destroyed, to prevent its interfering in any manner with the defence. A thick-set hedge, (Fig. 61,) if over six-and-a-half feet high, should be cut down to this height, and the cuttings be set into the hedge to render it less penetrable; a small ditch is dug in front of the hedge, the earth from which serves to form a banquette and a slight parapet, which are thrown up against the hedge. If the hedge is less than six-and-a-half feet high, (Fig. 62,) it is cut down to the height

of four-and-a-quarter feet; a ditch or trench, about three feet wide at bottom, and two feet deep, is dug behind the hedge, and the earth is thrown up against it, as in the last case. A width of two or three feet should be left between the trench and the earth thrown against the hedge to serve as a banquette.

243. A simple ditch behind a hedge will often serve as a good cover for light troops without any other preparation.

244. *Walls.* If a wall is of brick, or stone masonry, or of mud, and at least six-and-a-half feet high, it may be arranged for defence by cutting loop-holes through it, at three feet apart. For walls two feet thick, the interior width of the loop-holes should be fifteen inches, the exterior width four inches, and the interior height twelve inches, and the exterior should be arranged to allow the fire to sweep as great a space as practicable. The bottom of the loop-hole is four-and-a-quarter feet above the ground. A small ditch, about three feet in depth, should be dug on the outside of the wall, and the earth thrown against

it, to prevent the enemy from closing on the loop-holes.

245. If the wall is less than six-and-a-half feet high, (Fig. 64,) it should be cut down to the height of four-and-a-quarter feet; or else a banquette should be made against it. A trench is dug behind it, as in the case of a hedge of the same height; a ditch of about two feet deep should be dug in front of the wall, to prevent the enemy from reaching the assailed with the bayonet, and to place an obstacle to his climbing over the wall.

246. Walls which are eight feet high, (Fig. 65,) may be arranged with a double tier of fire; for this purpose a banquette of boards, sustained by *trestles*, or by casks, is made, to enable the assailed to fire over the wall; loop-holes are made near the foot of the wall, and a ditch is dug behind it, to obtain a second tier of fire. In this case no ditch should be made on the outside of the wall, because it might enable the enemy to close on the loop-holes.

247. If a long line of wall should require flank-

EXPLANATIONS OF PLATE IX.

Fig. 55. Shows the Profile on a hill side where the terre-plein is excavated behind the parapet.

Fig. 56. Shows a similar case to Fig. 55; the natural face of the hill is cut into steps to prevent the embankment from sliding.

Fig. 57. Shows a Profile of an Escarped hill side.

Fig. 58. Shows the plan of a work to dispute the passage of a ford.

Fig. 59. Shows the plan of a Tête-de-Pont, composed of a Priest-Cap A, with flanks, or crotchets; and flanking batteries C on the opposite shore.

Fig. 60. Shows the plan of a Tête-de-Pont, composed of a simple Crown A; a system of advanced lunettes D; and a flanking arrangement of batteries at D and C on the island, and opposite shore.

Figs. 61 and 62. Show Profiles of a Defensive Hedge.

Figs. 63, 64. Show Profiles of Defensive Walls.

Plate.

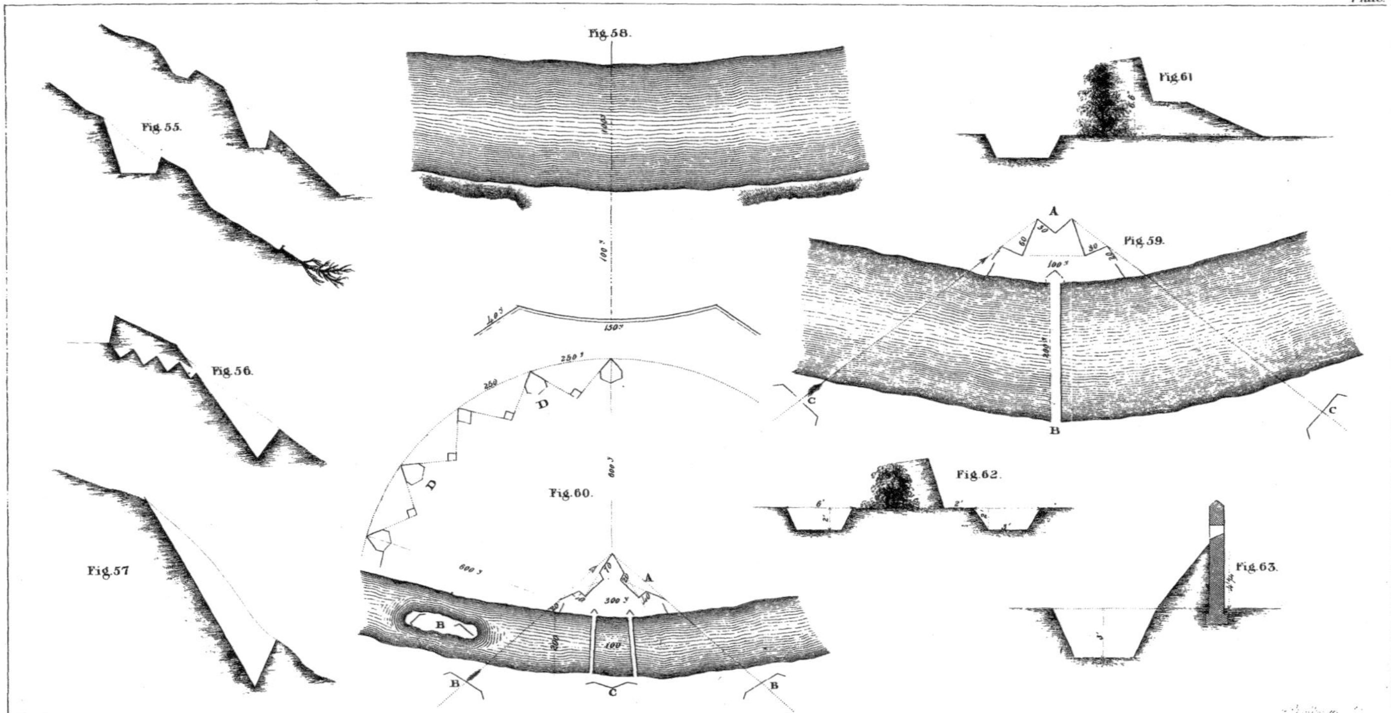
Fig. 55.
Fig. 56.
Fig. 57
Fig. 58.
100 y
150 y
Fig. 60.
250 y
250
D
D
600 y
600 y
300 y
100
200
A
B
B
B
C
Fig. 61
Fig. 59.
A
60
30
30
20
100 y
B
C
C
Fig. 62.
6'
2'
Fig. 63.

ing arrangements, a timber work, termed a *tambour*, is made on the outside; this work is made like a defensive stoccade, with square or round timber about six or eight inches in thickness; the plan of the tambour is that of a redan with a salient angle of 60°. A hole is made through the wall to communicate with the tambour, and loop-holes are made in the wall to flank the faces of the tambour.

248. *Barricades.* Wagons, or carts, sunk in the ground up to their axle-trees, and loaded with stones, or earth, will form a very good barricade. A pile of loose stones will serve the same purpose in some cases; also timber laid in piles, lengthwise and crosswise, with the open spaces filled with stones, or earth, forms an excellent barricade.

249. Barricades should be defended from behind with musketry, and cannon loaded with case shot; a work, like a defensive stoccade, will therefore be the best arrangment for them.

Houses. To place a house in a defensive attitude, the doors, and windows, of the lowest story should be firmly barricaded, and loop-holes be made as in the case of a wall. A tambour should be placed before the doors, both for their protection and to procure flanking arrangements if required. The windows of the upper stories should be partly barricaded, to cover the troops within, and loop-holes should be arranged as in the lower stories.

250. The roof, if not fire-proof, should be torn down, and the floor of the upper story be covered with earth or dung, moist from the stable, to the depth of about two feet.

251. If it is intended to defend the upper stories, should the enemy succeed in forcing the lower, the stairs should be torn down, and slight ladders be used in their stead; holes should be made through the floor to fire on the enemy in the lower story, or to throw weighty articles, or boiling water, &c., on him.

252. If there are balconies to the windows of the upper stories, or an upper gallery, they can readily be placed in a defensive attitude, by placing thick boards as a shelter on the outside,

and cutting holes through the floor to defend the doors and windows of the lower story.

253. If there are no conveniences of this nature, a temporary structure, termed a *machicoulis gallery*, (Fig. 67,) may be formed, by placing stout pieces of scantling through holes made in the wall, on a level with the floor; these pieces being confined to the floor on the inside, either by nailing them to it, or by tying them with rope to the joists; they should project from three to four feet beyond the wall on the exterior, and vertical pieces of smaller scantling, about four feet long, should be nailed to them, on which boards are nailed to cover the troops from the enemy's fire; these boards should be at least three inches thick. The flooring of the gallery is laid on the horizontal pieces, and holes are made through it to fire on the enemy, or to throw grenades, stones, &c., on him.

254. Any similar arrangement which will shelter a man, in the act of firing from a window on the foot of the wall, or in throwing over stones, &c., will serve the same purpose as a machi-

coulis gallery. A table might easily be arranged to answer the end in view.

255. In arranging a house for defence, every thing should be turned to account. The chimneys may be partly torn down on the interior to obtain bricks to barricade the windows. The flooiing boards may be used for the same purpose, and pillows, blankets, books, &c., be thrust in between the boards, nailed on the inside and outside of the windows. By tearing away some of the partitions, and taking up some of the joists of floors, heavy timber may be procured for the machiconlis galleries, for buttresses, for the lower barricades, for shores, &c.

256. The intelligence of the officer will readily suggest to him the uses to which the objects at hand may be applied.

257. *Positions and Intrenched Camps.* There is no talent more essential to an officer than that of seizing at a glance the strong and weak points of a position. This talent, known by the name of the *Military coup d'œil*, can be acquired alone by practice and study; for whatever may

be said of natural gifts, no apprehension, however quick it may be, can supply the places of these indispensable requisites in every art, and in no one are they more so than in the *Military Art*.

258. Positions derive their great importance from the influence of fire-arms in the decision of battles; for whatever enables one party to deliver its fire with effect against the other, whilst it, at the same time, remains sheltered in any degree from that of its adversary, places the advantage, all other things being equal, greatly on its side; and it is this advantage which should be principally kept in view in selecting a position.

259. Woods, commanding heights, precipices, and villages, constitute the strong points of a position. They serve as points of support against which the wings of the army rest; or else, by covering parts of the front, they serve as the key points in the defence.

260. A wood, if properly intrenched, covers the infantry from the attacks of cavalry; conceals its manœuvres, and enables it to deliver

its fire without being exposed to that of the enemy.

261. Heights, by giving a commanding view of the surrounding ground, increase both the range and the effects of fire arms; whilst they, at the same time, serve to screen the troops behind them until they are required to be brought into action. Precipices offer similar advantages to heights, and are moreover unassailable.

262. Villages serve as secure shelters for detachments, which, by their fire, cover the manœuvres of the troops in their rear; and, if properly intrenched, will cause the enemy great loss in his effort to force his way into them.

263. Rivers, marshes, hollows, and ravines are the most unfavorable features of a position because they prevent a free circulation from one point to another, and thus impede the manœvres; and they are exposed to the full fire of the enemy. They may however be of service when they are so placed as to support the wings.

264. The best positions are those which, being in due proportion to the force by which they are

occupied, command all the surrounding ground within cannon range, the ground descending in a gentle slope to the front, presenting woods, villages, &c., to support the wings, and cover parts of the front, and admitting of a free circulation from one point to another, with secure communications in their rear in case of retreat. If with these advantages, they present marshes, or other obstacles, which will embarrass the enemy's movements, and force him to advance in column, exposed to the fire and free manœuvres of the assailed, they will reunite every thing desirable in a favorable field of battle.

265. Troops should always be encamped in order of battle, to avoid the consequences of a surprise. When a camp is fortified, it is termed an *intrenched camp.* The same rules and general principles, apply to the choice of a site for a camp, and the manner of fortifying it, as to other positions. These rules and principles have been so fully developed in the preceding part of this work, that there remains nothing more to be said here which would not be a repetition of what has already been laid down.

CHAPTER XI.

ATTACK AND DEFENCE OF INTRENCHMENTS.

266. *Attack.* The subject of the attack admits of two natural divisions; the first of which comprehends all the preliminary steps taken before the troops are brought into action; the second all the subsequent operations of the troops.

267. An attack is made either by *surprise*, or *openly*. In both cases exact information should be obtained of the approaches to the works; their strength; the number and character of the garrison; and also the character of the commander. This information may be obtained through spies, deserters, prisoners, and others who have access to the works; but implicit faith ought not to be placed in the relations of such persons, as they may be in the interests of the garrison; and in all cases they should be strictly cross-examined

and their different representations be carefully compared with each other.

268. The only source of correct information is an examination, or *reconnaissance*, made by one or more intelligent officers. This reconnaissance should, if possible, be made secretly; but as this will not be practicable if the garrison show even ordinary vigilance, it will then be necessary to protect the reconnoitering officer by small detachments, who drive in the outposts of the garrison. The object to be attained by the reconnaissance is an accurate knowledge of the natural features of the ground exterior to the works; the obstacles it presents, and the shelters it affords to troops advancing; the obstacles in front of the counterscarp and in the ditches; the weak and strong points of the works, and the interior arrangements for the defence. If the work is an isolated post, information should be obtained as to the probability of its being succoured in case of an attack; the length of time it must hold out to receive succour; and the means it possesses of holding out.

269. *Attack by surprise.* A surprise is an *unexpected attack*, for which the assailed are not prepared. It is, perhaps, the best method of assailing an undisciplined and careless garrison, for its suddenness will disconcert and cause irremediable confusion.

270. Secrecy is the soul of an enterprise of this nature. To ensure it, the garrison, if aware of the presence of the troops, should be deceived and lulled into security by false manœvres. The troops that form the expedition should be kept in profound ignorance of its object until they are all assembled at the point from which they are to proceed to the attack.

271. The winter season is the most favorable for conducting a surprise, which should be made about two hours before day, as this is the moment when the sentries are generally least vigilant, and the garrison is in profound sleep; and the attempt, if at first successful, will be facilitated by the approach of day, and if unsuccessful, the troops can withdraw with safety under the obscurity of night. Should there be danger, from suc-

cours arriving in a short time, the attack should be made soon after midnight, when the garrison is asleep, so that the troops may retire before daylight, after having attained their object.

272. The troops for the attack should consist of a storming party and a reserve of picked men. There should be two guides, one in front of the storming party, with the detachment of workmen under the command of an engineer officer; the other in the rear, under charge of a guard, to supply the place of the first, if killed. The workmen should be furnished with axes, crowbars, pick-axes, &c., and several bags of powder, of about thirty pounds each, to be attached to palisadings, fraises and barriers, to blow them down, if the alarm should be given whilst they are opening a way through them by other means. All the operations should be carried on with despatch and in silence. Should the sentries challenge, they must be secured or bayonetted.

273. Circumstances alone can determine whether it will be advisable to make false attacks with the true one. They will distract the atten-

tion of the garrison, if the alarm is given, from the true attack; and a false attack has sometimes succeeded when the true one has failed. When made, one should be directed against the strongest point of the work; as the strong points are usually guarded with less vigilance than the others; and they should all be made at some distance from the true attack; and orders be given to the detachments making them to proceed to the point of the true attack, should they succeed in making their way into the work.

If the attack succeeds, immediate measures should be taken to place the works in a state of defence, if the position is to be maintained; or else they should be destroyed, as far as practicable, before retreating from them.

274. *Attack by Open Force.* The general arrangements for an open assault, comprehend the operations to gain possession of the works; the measures for maintaining possession of them, and following up the first advantage; and, finally, the precautions to be observed in the event of a repulse.

275. An open assault may be made either with the bayonet alone, or with the combined action of artillery and the bayonet. The first is the most expeditious method, but it is attended with great destruction of life; it should therefore only be resorted to against works of a weak character, which are feebly guarded; or against isolated posts within reach of speedy succour. When resorted to, it will usually be best to make the attack just before day. If it is made by daylight, it will, in most cases, be well to scour the environs with a few squadrons of cavalry, to pick up patroles and stragglers who might give the alarm; and then push forward rapidly the assaulting columns. If the assailed seem prepared, light troops should precede the columns of attack, with orders to display in front of the counterscarp, and open a brisk fire on the assailed, for the purpose of diverting their attention from the columns of attack.

276. In an attack with artillery, the troops are drawn up in a sheltered position, or beyond the range of the guns of the assailed; batteries are

then established within about six hundred yards of the works, in the most favorable positions to enfilade the faces, and destroy all visible obstacles. The batteries keep up an incessant fire of ball and hollow projectiles, in order to dismount the cannon, and create confusion among the assailed. When the fire of the works is silenced, the troops are brought forward, and demonstrations are made on several points, to divert the attention of the assailed from the true point of attack, and prevent him from concentrating his strength on that point. Several false attacks should be made at the same moment with the real one, and each of them should be sufficiently formidable, in point of numbers, to enable the troops to profit by any success they may obtain.

277. The number and disposition of the troops making the assault will depend, in so great a degree, on local circumstances, and the arrangements of the assailed, that nothing more can be laid down under this head than some general rules.

278. The attack should be led by a storming

party, composed of picked troops, or of volunteers for the occasion; this party is preceded by a detachment of engineer troops, provided with the necessary means to make their way through all obstacles, to enable the storming party to assault with the bayonet. If the detachment is arrested at the crest of the counterscarp, by obstacles which must be destroyed before farther progress can be made, the storming party may display, and open a fire on the assailed to divert their fire from the workmen; but this operation should only be resorted to from necessity, as it breaks in on that unity so essential in an operation of this character, and impairs the confidence of the soldier in the bayonet, on which his sole reliance should be placed.

279. The storming party should be provided with light scaling ladders, planks, fascines, strong hurdles, &c., for the purpose of descending into the ditch; to mount the scarp; to cover trous-de-loup, small pickets, &c., &c.

280. Another detachment of engineer troops follows in the rear of the storming party; its

duties consist rendering the passages, opened by the first detachment, more accessible to the troops, who immediately follow it to sustain the storming party. This second detachment is also charged with the care of placing the work from which the assailed has been driven in a defensive attitude, in order to frustrate his attempt to repossess himself of it. The first detachment cannot be charged with this duty: for, independently of having handsomely acquitted itself in bearing the brunt of the action, it may be required to precede the storming party in the pursuit of the assailed to his interior works.

281. The troops destined to support and, if necessary, reinforce the storming party, advance in one or two lines, with cavalry, and some pieces of artillery on the wings, to repel sorties. The remainder of the troops follow in order of battle, to improve the first successes, or to cover the retreat of the assaulting columns, if repulsed.

282. The salients are generally the points on which the storming party advances, unless some natural feature of the ground should present

greater facilities for advancing on a re-entering, or in front of a face. When the ditch is gained, shelter is sought in a dead angle; and, if the work is fraised, or resolutely defended with the bayonet, a breach must be made, either by firing beforehand hollow loaded projectiles into the parapet, or by undermining the scarp with the pick. If the intrenchments consist of detached lunettes, an attack should be made on their gorge at the same moment with the one in front.

283. When the assailed are driven from their works, the storming party should press hotly on their rear, and endeavor to enter pell-mell with them into their interior works, leaving to the troops which follow them the care of retaining possession of the works gained; but, if the intrenchments are supported by other troops drawn up in order of battle, the storming party should halt in the works until it is reinforced by the troops in its rear.

284. There is no danger to be apprehended, in case of a retreat, after an unsuccessful attack on

an isolated work. But in an attack on intrenchments, supported by an army, the retreat of the storming party should be covered by cavalry and artillery, until it can find safety behind the main body of the troops, drawn up in order of battle to protect it, and to receive the assault of the assailed, should he attempt offensive operations.

285. In conducting the attack, preparations should be made beforehand, to remove all the artificial obstacles that the assailed may have placed before their works, to impede the progress of the storming party. This duty is usually intrusted to the detachment of engineer troops, who are provided with axes, picks, and other suitable tools for this purpose. When the obstacles are of a nature to be easily destroyed by artillery, it should always be resorted to. Abattis, palisades, fraises, and entanglements when exposed, may be readily torn to pieces by opening an enfilading ricochet fire on them. Trous-de-loup may be passed, either by covering them with plank or strong hurdles, or else by directing the men to pass slowly between them in extended order; small pickets may be

broken down, or else fascines may be spread over them so as to form a tolerably stable road-way; fascines may also be used to cover the points of spikes.

286. The passage of the ditch, and the assault of the parapet, are the most difficult operations. If the ditch is not more than six feet deep, it can be leapt into without danger, and the men can mount the scarp readily with a very little assistance from each other. When the width is not greater than twelve feet, the ditch may be crossed by laying thick plank, or small scantling, over it. When the depth is over six feet, the storming party is usually provided, either with small scaling ladders, or with fascines, sand-bags, or other means, to fill the ditch partly up; or if these means cannot be procured, the detachment dig away the counterscarp, throwing the earth into the ditch, and thus, in a short time, form an easy entrance to it. Wet ditches may be filled up either with sand-bags alone, or more expeditiously by attaching sand-bags to large fascines, or to trusses of hay or straw.

287. The assault of the parapet is made by the aid of scaling ladders, or by effecting a breach, by firing loaded hollow projectiles into the scarp and parapet, which, by their explosion, crumble the earth down, so as to form an accessible ramp; or else the foot of the scarp is undermined with the pick, and the mass of the parapet falls into the ditch.

288. *Defence.* The essential point in the defence is to have every part of the works guarded by a sufficient number of troops to resist an attack on all sides. This is of importance not only in isolated works, which can be surrounded, but also in continued lines; for although the enemy will usually make an attempt to enter only at a few points, still the confusion and delay, which might arise from changing the position of the troops, to meet the movements of the enemy, particularly in a night attack, would be extremely hazardous.

289. A vigorous defence will require, at least, two ranks to be drawn up on the banquette throughout the entire extent of the line, with

a reserve proportioned to the importance of the work.

290. The strictest vigilance should be exerted to guard against a surprise; for this purpose sentries should be posted on all of the most commanding points of the works; and on the exterior, at such points as the enemy might approach, as roads, fords, defiles, bridges, &c.; besides these, small detachments of picked men should be stationed to watch these accessible points, and all other places where the enemy might secrete himself, or approach unexposed to the fire of the works. Patroles should be sent out to watch the enemy's movements, and to have an eye on the manner in which the sentries perform their duties. The patroles are particularly charged with the duty of preventing the enemy from reconnoitering the approaches to the works; to do which effectually, without giving unnecessary alarms, they should be ordered to attack with the bayonet alone, unless surprised by an ambuscade.

291. At night the number of sentries should be increased; and redoubled vigilance be used, par-

ticularly after midnight. If there are well grounded apprehensions of an attack, *fire-balls* may be thrown out, or fires be lighted in front of the works to discover the approach of the enemy. Occasional false alarms should be made to keep the garrison on the alert; but this artifice should not be too often practised, otherwise it might produce a contrary effect; besides all useless fatigue should be spared men who are sufficiently harassed by the ordinary duties of their situation.

292. In conducting the defence, each corps should have its particular post assigned to it; and receive special instructions relative to the part it will have to play. The reserve is posted in the most convenient position to afford prompt assistance to any point in danger of being forced. The troops drawn up on the banquette should be carefully instructed in their duties; and they should be taught not to look for aid, or relief, from the reserve, until they have repulsed the enemy.

293. If the enemy opens his attack by a warm cannonade, the troops should not be exposed to

it, if they can be sheltered near the posts they are to occupy when his columns of attack approach, at which time his cannonade must cease, in order not to injure his own troops. The men should be instructed to reserve their fire until the enemy has arrived at certain points marked out in front of the works, which should not be farther than two hundred yards from the parapet.

294. Sorties, if well timed, will generally decide the fate of the affair. They should be made when the column is thrown into confusion, or shows any signs of irresolution in its movements; or when its flank, or rear, is carelessly exposed. They are sometimes made to alarm the enemy for his own safety; or to make a diversion in favor of expected succours in blockaded posts.

295. Cavalry is the best arm for a sortie, on account of the rapidity of its movements, and the violence of its shock. When infantry is employed for this purpose, it should use the bayonet alone.

296. Should the enemy succeed in forcing his

way into the work, the reserve should attack with the bayonet, before he has time to form; but it must be confessed that success oftener crowns an offensive movement, on the part of the assailed, in endeavoring to regain possession of their works, than any effort to drive back the enemy at the moment, when flushed with success, he has the hope of a certain victory. The only well-grounded prospect that the assailed have of repelling the assault, when the enemy has gained the top of the scarp, is to meet him in an offensive attitude at the point of the bayonet on top of the parapet.

297. The particular arrangements of the defence consist in defending all obstacles, such as abattis, palisadings, &c., by a warm well-aimed fire; as the particular object of these obstacles is to keep the enemy exposed for a longer time to the fire.

298. Large stones, heavy round logs, and loaded hollow projectiles, should be in readiness to be rolled over on the enemy whilst he is in the ditch endeavoring to mount the scarp. Large branches

of trees prepared as for an abattis, with chevaux-de-frise, and other obstacles, should be at hand to obstruct the breach.

299. Finally, in an isolated post, if the enemy, after having been repulsed, makes a show of blockading it, or of renewing his attack, and there is no prospect of succour arriving, the garrison should attempt an escape by night.

END OF THE FIELD FORTIFICATION.

CHAPTER XII.

MILITARY COMMUNICATIONS.

300. DESPATCH is of the first importance in all military operations; and nothing contributes more to it than the means of opening practicable and easy communications in every direction.

301. Among the various obstacles by which the free movements of an army may be embarrassed, water courses, marshes and swamps, hold the first rank, owing to the difficulty of finding suitable means at hand to pass them, or of transporting those means in the train of an army.

302. For the passage of water courses, bridges must be constructed; the character of the construction depending on the nature of the river, the particular locality selected for the bridge, and the means at hand.

303. Before adopting definitively the particu-

lar locality, the officer charged with the construction should endeavor to gain the most accurate information, on every point, which might in any manner bear upon this selection; for which purpose he will cause a reconnaissance to be made, to determine the breadth and depth of the river, and the character of its bed;—the velocity of the current, the seasons of freshets, with their usual extent and effects;—the character of the banks, their absolute and relative heights, their slope, the approaches to them, the distance between them and the water way, and finally, if they are wooded or otherwise;—the position and size of islands, whether wooded or not;—the extent of fords, their depth of water, and the nature of their bottom;—the form and dimensions of elbows and indentations;—the artificial obstructions, or dams, &c., the possibility of their being destroyed by the enemy, and the practicability of applying them to the purposes of artificial inundations;—the position, extent, and character of the affluents;—a description of the permanent bridges, and ferries;—whether the river empties itself into the sea, or into another river, the extent and effects of the tide water;—the nature of the permanent and temporary defences along the river;—the kind of boats used in navigating the river, and the time it would require to collect them at a given point;—the proximity of forests to its banks, and the facilities afforded by them for constructing rafts.

304. As a general rule, the elbows, which present their concavity to the side from which the passage is to be made, are the best positions for the bridge, as being most favorable to the defence; but it must be observed that, when the bridge is to remain any length of time, the effects of the current on the banks, at the elbow, should be carefully ascertained; these effects usually consist in washing away the banks on the concave side, and transporting the fragments to the opposite shore, thus deepening the bed on that side, and filling it up on the other.

305. *Military Bridges.* Any structure raised especially for the purpose of affording a passage to troops across a river, may be termed a *military*

bridge; but the expression is generally restricted to those temporary expedients for the same end, of whatever character they may be, to which an army is obliged to resort during the active operations of a campaign.

306. The means most frequently resorted to, for the passage of troops over a river, is a *floating bridge.* This construction consists of boats, rafts, casks, or other buoyant bodies, anchored at suitable distances apart, to serve as points of support to the bridging proper, which is formed of beams, resting on the supports, to which they are firmly connected, as well as with each other, and of a flooring of plank laid across the beams, and fastened to them.

307. Any buoyant body may be used for the supports, but experience is in favor of large flat-bottomed boats, termed *bateaux,* as presenting more advantages than any other. In Europe, bateaux termed *pontoons,* which consisted of a frame work of light wood, covered on the bottom and sides with tin, or copper, were for a long period mostly in use as a part of the bridge train

of an army; but they have gradually gone out of use, and the French particularly, whose corps for the construction of military bridges, termed the *pontooneers*, is justly held in the highest estimation, as the most scientific and experienced in Europe, have adopted the bateau of uniform dimensions, with corresponding equipments, for its army boat train.

308. As it is not the object of this chapter to give a full account of the subject of military bridges, which more properly belongs to a separate treatise, it will be confined to describing only such as troops, the least skilled, may construct under the guidance of an intelligent officer.

309. The general properties of all bridges should be solidity, stability, and, if they are to be transported on land, lightness.

310. The solidity will depend on the nature of the points of support; and on the cross section of the beams, and their *bearing*, or distance, between their points of support.

The solidity should be such as to resist the action of the heaviest loads on the bridge, and also that arising from the current, the waves, and the wind.

311. The essential requisite for the solidity of a floating bridge is, that it shall not be loaded with a weight sufficient to submerge it. To ascertain the greatest weight that a bridge may be required to bear, the following data may be taken. The average weight of a foot soldier, with his arms, may be assumed at two hundred pounds; and the lineal space occupied by each man, when marching by a flank, eighteen inches; a cavalry soldier, with his horse and equipments, may be taken at fourteen hundred pounds, and he occupies about nine lineal feet in file; a twelve-pounder gun with its carriage, limber, six horses, and three drivers, weighs about eleven thousand six hundred pounds, and occupies about forty lineal feet in march. The weights here given are rather over estimated, so that any error may be on the safe side.

312. The term *bay* is applied to the interval between any two points of supports of a bridge, estimating this distance between the middle

points of the supports; for example, in bateau supports, it would be from the centre line of one bateau to the centre line of the other. If we suppose this bay to be forty lineal feet, a very simple calculation will show, that when infantry is marching by a flank, in the ordinary formation of three ranks, there will be about eighty men on the bay, or sixteen thousand pounds in weight; and comparing this with the weight of either artillery or cavalry, it will be found to exceed them both. It may therefore be assumed as the greatest weight that the bridge will have to bear on each bay.

313. By a principle of hydrostatics, the weight of the volume of water displaced by a floating body is equal to the weight of the body itself; if, therefore, the cubic contents of the part of the bateau, which is above the water line after the bridge is finished, be estimated, and this volume be multiplied by sixty-two-and-a-half pounds, the weight of a cubic foot of water, the result will show the weight, which in addition to that of the bridge, would submerge a bateau; and as the

weight, placed on each bay, is borne by the half of each bateau of the bay, it follows, that if a weight, thus obtained, were laid on each bay, it would submerge the entire bridge. Now, in practice, it is found that a bridge cannot bear with safety a weight greater than three-fourths or four-fifths of that which would submerge it. This consequently will furnish a ready means of comparing the weight that may be placed on a bridge with what it can bear with safety.

314. To secure the stability of the bridge, every precaution should be taken to prevent any tendency to an oscillating, or undulating motion, in any direction. These precautions will, moreover, add to the solidity of the bridge, as the connection of the parts will be firmer, and the weight consequently more uniformly distributed over the supports.

315. Oscillations in a lateral direction are prevented by anchors placed above and below the bridge, to which the supports are moored. The object of the up-stream anchors is, principally, to resist the action of the current, and of those down-

stream to resist that of the winds and waves. The number, in each case, will therefore depend on the force of the current and on the width of the river, and its exposure to high winds.

316. The inclinations of the cables, by which the supports are attached to the anchors, should be the same for all; so that they may offer a uniform resistance to a pitching motion. This inclination is, ordinarily, taken one-tenth; that is, the distance of the anchor, from the support, is ten times the depth of the water. When the width of the river, and the nature of its banks, render it practicable, a cable is stretched across it, by means of a windlass, and the supports are either attached to this cable, by separate fastenings, or else the cables which attach them to the anchors are passed over and fastened to it.

317. Undulations in a longitudinal direction are prevented by connecting the beams firmly to each other, and to the points of support. This connection will also increase the solidity, by diffusing the weight thrown on one support over several. The manner of regulating the passage

of troops will, in a great degree, contribute in preventing this motion. Infantry should always pass in open order, and with the route step; cavalry should dismount and lead their horses; and artillery, in some cases, should be drawn over by men.

318. Lightness is an essential quality in the bridge equipage which is to follow an army, but it should always be subordinate to solidity and stability.

319. *Bateau Bridge.* As the bridge equipage of the French is considered the best arranged, a description of the bridge itself will be here given, to convey a more complete idea of this kind of structure.

320. The bateau is thirty-one feet long, two feet six inches deep, five feet four inches wide at top, and four feet wide at bottom. The bateau is built something like a *skiff*, the head and stern being a few inches higher than the body; the width at the head being two feet, and at the stern four feet. The ribs are of oak; the bottom and sides of pine plank one inch thick.

321. The beams, termed *balks*, which are laid

across the bateaux, are of four-and-a-half inch scantling, and twenty-five-and-a-half feet long.

322. The flooring of the bridge is formed of boards, termed *chesses*, cut in lengths of fourteen feet, the width being twelve inches, and thickness, one-and-a-half inches.

323. The bateaux, when the bridge is formed, are thirteen feet ten inches apart, making the distance between their centre lines, nineteen feet two inches.

324. Each bay has five balks, the centre lines of the balks being two feet ten inches apart; and the ends of the balks project six inches beyond the outside gunwales of the two bateaux on which they rest.

325. To confine the chesses to the balks, pieces of scantling, termed *side-rails*, are placed on them, over the outside balks, to which they are fastened by a cord, by what is termed a *rack-lashing*.

326. The distance between the side rails is about eleven feet, and is the width of the free

road-way. The chesses project equally beyond the outside balks.

327. Before the construction of the bridge is commenced, a road-way of easy access, the slope of which may be only one-sixth, should be cut through the banks if necessary. An *abutment* is then formed, by laying a stout beam, in a horizontal position, perpendicular to the direction of the bridge. The length of this beam should be the same as the chesses; it is imbedded in the earth, on the same level as the top of the bridge, and is confined in its place by a stout picket driven at each end of it, and by two others driven in front of it. A chess is laid flat behind the beam, and with it constitutes the abutment.

328. To give the proper alignment to the bridge, a staff is placed at a short distance from the abutment, in the direction of the centre of the bridge, and a second staff is placed some yards in the rear of the first, and on the same line.

329. The bateaux are placed in the water below the point which the bridge is to occupy; and they

are brought successively into their proper position, as the bridge is carried forward from the shore. The first bateau is placed, when the depth of water will admit of it, so far from the abutment that the balks resting with one of their extremities on it, the others will project six inches beyond the outside gunwale of the bateau. The positions of the outside balks are shown by *pintles* placed in the gunwales. The chesses are laid on within one foot of the first bateau; the position of the second bateau is then determined, and the balks of the next bay are laid; the precaution being taken to lay all the balks of the second bay either on the up-stream, or the down-stream side of those of the first bay. The chesses of the second bay are then laid, within one foot of the second bateau, and the construction is thus regularly carried on.

330. As the balks are laid they are fastened to each other, and also to the bateaux, by a rack-lashing; iron *crotchets* being placed on the inside of the bateau for this purpose.

331. The bateaux are fastened to each other by

two *spring lines*, which pass diagonally from the head of one bateau to the stern of the other; they pass through iron rings placed in suitable positions on the outside of the bateaux. The four bateaux, nearest the shore, are fastened to strong stakes, on shore, by two ropes for each bateau; one up-stream, the other down. An anchor is allowed up-stream to two bateaux, and one down-stream to four.

332. When a cable, or *sheer line*, can be stretched across the river, before commencing the bridge, the preceding operations will be greatly expedited; each bateau being attached at once to the cable, by ropes termed *head-lashings*.

333. The side rails are laid so soon as the third bay is commenced; they are fastened to the outside balks by a rack-lashing; the chesses being so arranged as to admit its being passed between them. Each rail should be fastened at its middle point, and near each of its ends.

334. When the flooring is completed, it is covered either with a thin layer of straw, or sand, to protect it in some degree from wear and tear.

335. The bridge here described underwent a series of rigid experiments to test its properties; the results of which were of the most satisfactory character, as to its solidity, stability and lightness.

336. When the bridge was completed the draught of the bateaux was nine-and-a-half inches; when traversed by a column of infantry three abreast, marching with the cadenced step, and in close order, the draught of the bateau was one foot seven inches; and was the same when crossed by a twenty-four-pounder, drawn by eight horses, with their drivers.

337. Its travelling equipage can follow in the train of an army with the same ease as a twelve-pounder.

338. A bridge across a navigable stream should admit a free passage to the river craft; for this purpose a part of the bridge, resting on one or two bateaux, should be so arranged that it can be slipped out of its place, forming a *draw* for the passage.

339. To break up the bridge, the operations are

EXPLANATIONS OF PLATE X.

Fig. 65. Shows a section and an elevation of a defensive wall with two tiers of fire; c, c′ is a platform, on trestles a, a′ for firing over the top, with a wooden ramp b, b′ to ascend to it; d, d′, are the lower loop-holes with a ditch behind the wall to admit firing through them.

Fig. 66. Shows a plan of a defensive wall A; flanked by a Tambour B.

Fig. 67. Shows a section of the wall of a house through a window, and a Machiconlis; a are horizontal pieces on a level with the floor, fastened to it by a horizontal cross-piece b, or by lashing them to the joists d; c are uprights fastened to the cross-pieces on which thick plank are nailed to form a parapet.

Fig. 68. Shows a plan, a longitudinal, and a cross section of a Bateau Bridge; A are the bateaux; B the Balks; C the Chesses; D the Side Rails; E the Abutment; F the Cross Lashing; S stakes to which the shore bateaux are moored.

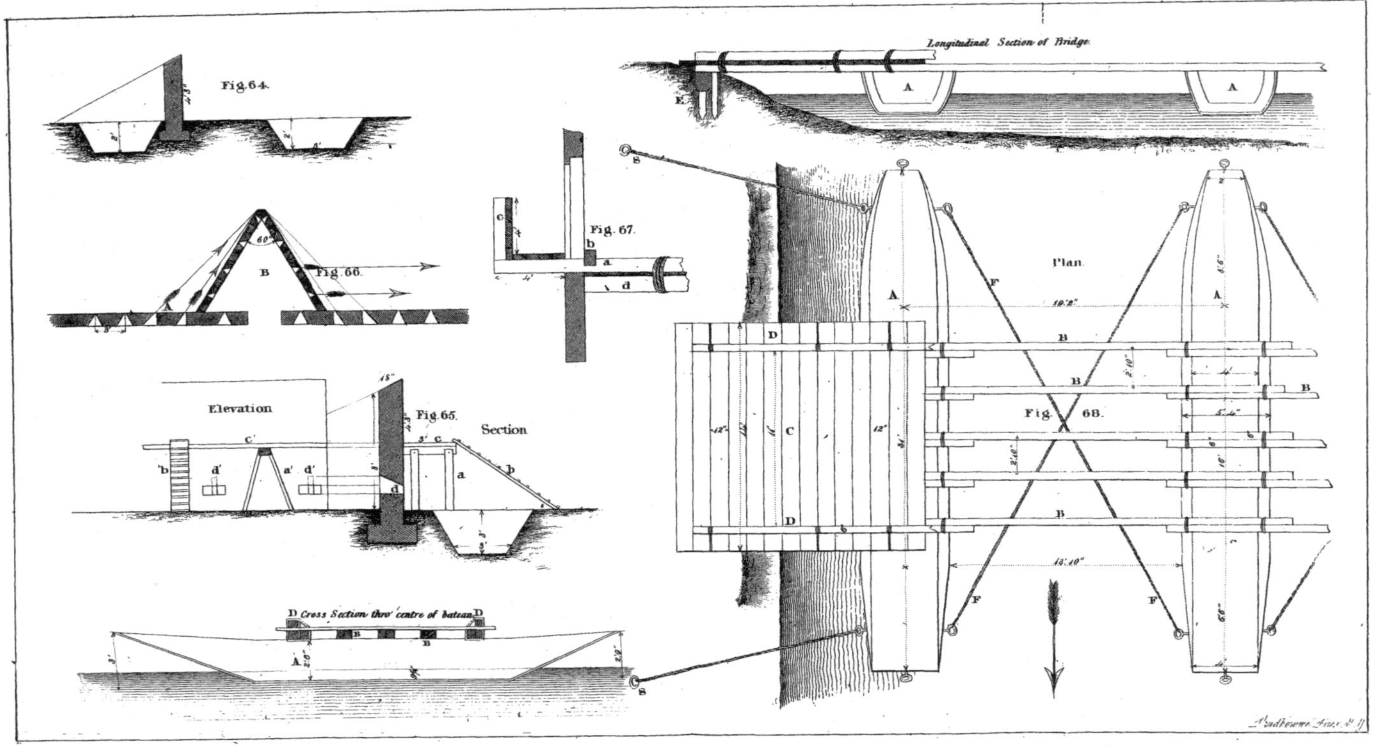
Fig. 64.
Fig. 66.
Fig. 67.
Elevation
Fig. 65.
Section
D Cross Section thro' centre of bateau D
Longitudinal Section of Bridge
Plan
Fig. 68.

the reverse of those for laying it. The operation is commenced at the shore to be abandoned; the side rails, chesses, balks, &c., are taken up in their proper order, and conveyed to the opposite shore, either by hand, or in the bateaux.

340. On narrow rivers, with a gentle current, this operation can be greatly expedited, by detaching the entire bridge from its moorings, and allowing it to swing round against the shore. To perform this manœuvre with safety, two stout ropes, or *guys*, are attached, one to the stern, the other to the head of the second bateau, from the shore to be abandoned, ten or twelve men are placed at each guy, to ease off, or *belay*, according to circumstances. Another stout rope is attached to the fourth or fifth bateau, from the opposite shore, and is drawn tight and fastened to a strong stake up-stream, at fifty or sixty yards from the bridge. When these preparations are made, the head lashings are cast loose; the two extreme bateaux are detached from their abutments; and the anchor cables are gradually

eased off, to allow the bridge to swing round with a gentle uniform motion.

341. This manœuvre requires great care, and should not be resorted to, except in cases of necessity, unless the current is gentle, and there are not more than twenty-five or thirty bateaux. The same means may be taken, under similar circumstances, to throw a bridge across a river.

342. *Boat Bridges.* Flat-bottomed boats are the most suitable for a bridge, owing to their greater volume, and their stability in the water; but when none other can be obtained, keel boats may be taken. If the gunwales are not suitable, to serve as points of support to the balks, a frame work should be placed in the centre of the boat for this purpose. The frame work consists of a sill-piece of six or eight-inch scantling, fitted to the bottom of the boat, and a cap-sill of the same dimensions, which is supported at a suitable height by uprights of four-inch scantling mortised into the cap and sill pieces. The whole is kept steady by *struts*, of the same sized scant-

ling as the uprights, which are placed against the uprights and ribs of the boat.

343. The balks are laid on the caps, jutting about two feet beyond them, and they are firmly secured to them and to each other by a rack-lashing.

344. As the boats are generally of unequal size, two small boats are, in some cases, lashed together to form one support; and to bring them all to the same level, they may be ballasted with stone. The strongest boats should be placed at the extremities of the bridge, as these parts are more liable to injury, from the motion of the bridge, than the middle.

345. When anchors cannot be procured, mill-stones, strong baskets, or bags, filled with large round pebbles, or broken stone, may supply their places; and, in shallow rivers, each boat may be fastened to stout stakes driven into the bed of the river.

346. *Raft Bridge.* Rafts formed of solid timber, or of casks, barrels, the skins of animals prepared and inflated, &c., are frequently used in military operations, either to construct bridges,

or to serve the place of boats in the transportation of troops across streams.

347. Timber rafts are particularly serviceable in wooded and mountainous countries, where, owing to the nature of the streams, boats cannot always be procured. A bridge may be constructed on rafts over rivers of any size, and may be made to bear any loads; but, owing to the form and dimensions of the rafts, rapid currents present great difficulties to their manœuvre whilst the bridge is under construction.

348. The timber for a raft may be either square or round, seasoned or green; in all cases, however, the lightest and best seasoned wood should be preferred as being most buoyant.

349. To ascertain the weight which will submerge a piece of timber of given dimensions, the following method may be used: Take a cubical block, of any size, and ascertain its weight; fill a vessel with water, and immersing the block in it, find the weight of the water which it displaces; the difference between the weight of the block and that of the water displaced, will give the weight

which, placed on the block, would be sufficient to submerge it. If this difference be multiplied by the weight of a cubic foot of water, and the product be divided by the weight of the water displaced, the quotient will give the weight necessary to submerge a block of a cubic foot; therefore, multiplying this result by the number of cubic feet in the given piece, the product will be the total weight sought.

350. The cubic contents of the piece itself, if it be the trunk of a tree, may be found by adding into one sum the areas of the two ends, and four times the area of the section, at the middle of the trunk, and multiplying this sum by one sixth the entire length.

351. The weight of green wood is greatest when it is cut whilst the sap is running; and seasoned wood, when placed in water, is found to augment about one-sixth in weight, in a few days, unless the precaution is taken to tar the ends of the trunks; and this increase of weight is found to be greater in wood that has been long cut than in new timber.

352. The volume of the raft should be in proportion to the weight of the load that the bridge must bear; and the length and breadth, for the same volume, will be determined from the consideration, that the stability varies directly as the length, and inversely as the breadth; therefore, the longest and largest trunks should be selected, since fewer of them will be required for the same load, and the length will thus be increased at the expense of the breadth. The length of the trunks should not be less than forty or forty-five feet, particularly if the current is rapid, otherwise there will be a very sensible oscillating motion of the bridge.

353. Before constructing the raft, (Fig. 69,) the trunks are placed in the water, as they will there assume their natural position of stability, and they can be more easily moved about in it. The large ends of the trunks are cut away on their under side, like the shape of a child's whistle. The trunks are then laid side by side with their butt ends together, leaving about four inches between them, and so that they will project a little beyond each

other from the middle to the outside trunks, making an angle, of the perpendicular of which is one-fourth the base. The trunks are then united by four cross-pieces, one placed near each end, and two near the middle, about ten feet apart; the cross-pieces being fastened to the trunks, either by strong wooden pins driven into augur holes made through the pieces, or else by a strong rope-lashing. Three sleepers are laid across the centre cross-pieces, one over the middle of the raft, and the other two near the outside trunks; the object of these sleepers is to give the flooring sufficient elevation above the surface of the water to prevent the waves from breaking over it.

354. The rafts are anchored in the stream by two anchors for each, one up-stream, the other down, the cables of which should be drawn tight to prevent any pitching motion. The rafts should be placed as far asunder as practicable, to allow a free passage to the water; this distance will depend chiefly on the size of the timber that can be procured for balks; each balk should jut at least

one foot beyond the middle sleepers of the two rafts on which it rests. The balks are secured to the sleepers and to each other by rack-lashings. The rafts should moreover be connected by a beam at the head and stern.

355. The tendency of the current on the up-stream cables is to submerge the head of the raft; this tendency will be in part counteracted by placing the butt ends up-stream, and by the manner of arranging these ends; as a farther precaution, the flooring should be placed lower down-stream than the centre of gravity of the rafts, so that its weight may tend to raise the head of the rafts.

356. On navigable rivers, a draw must be made similar to the one used for a boat bridge; but as a *raft-draw* would not be easily manœuvred, particularly in a strong current, it would be well to use a draw of boats.

357. Rafts of empty casks, or inflated skins, may be made to bear any load, by lashing a sufficient number together. They cannot be relied

on for a safe passage except on narrow and gentle streams.

358. *Flying Bridges.* Any floating body which is propelled from one shore to the other by the force of the current alone, acting obliquely to its side, is termed a *flying-bridge.* But in order that the flying-bridge shall cross and recross always at the same points of the shores, it must be held by a cable, attached to an anchor, either in the middle, or at some more suitable point of the stream.

359. A flying-bridge, when arranged in the best manner, should bear the heaviest loads that the service may require; it should be perfectly stable, and cross the stream in the shortest time possible.

360. The most usual manner of forming a flying-bridge is to connect two boats, by a platform, consisting of joists laid across the boats, and a flooring of plank on the joists. The cable is attached to this platform, and passing over a cross-piece, supported above the platform by two uprights, so as to sweep clear of the deck in the

movement of the bridge, is secured at the other end to an anchor, or any other fixed point of support.

361. The best form of boats is one with a long, narrow, flat bottom and sharp head; the length from sixty to ninety feet; width, ten to fifteen feet, and depth, six to seven feet. These boats have great capacity, and present a large side surface to the action of the current.

362. The distance between the centres of the boats cannot be much greater than twenty-five to thirty feet, owing to the usual dimensions of the beams used for joists, which will not generally admit of a greater bearing. The length and breadth of the platform are about the same, and may be about fifty feet.

363. The bridge may have two platforms, like the double decks of steamboats, and when arranged in this way, by proper management, it may be made to bear from one thousand to fourteen hundred infantry.

364. The time of crossing will depend on the length of the cable, the position of the anchor,

and the direction given to the head of the boat, by means of a rudder, whilst in motion.

365. The cable should be at least equal to the width of the river, and experience has shown that the most suitable length for a quick passage, is once-and-a-half the width. If the velocity of the current is the same on both shores, the anchor should occupy the middle point of the stream; but if it is greater near one than the other, then it is best to place the anchor nearest to the shore where the velocity is the least, because in approaching that shore the position of the boat will be in favor of the lessened action of the current.

366. Ordinarily one anchor is deemed sufficient; but as the action of the boat will tend to shift the position of its *shank*, and consequently weaken the hold of its *flukes*, it would be better to place three together, the shank of the middle one being in the direction of the current, and those of the other two being respectively in the directions of the extreme positions of the cable.

367. The height of the cross-piece, which sup-

ports the cable, will depend on the velocity of the stream, and the length of the cable; its object being to keep the cable clear of the platform. A height varying between twelve and thirty feet is generally sufficient. The best position of the cross-piece and its supports, is about one-third the length of the boat from the bow. One cross-piece is enough to support the cable; but as the motion of the boat would occasion rapid wear and tear, both of the cable and the cross-piece, a better arrangement consists in placing two cross-pieces, with a sliding pulley, or *block*, between them, the cable passing through an eye made in the pulley.

368. A windlass is placed on the platform, towards the stern, to which the cable is fastened; it serves to let the cable out in case of danger, from the action of the current or wind, and to take up the *slack* when necessary.

369. In shallow gentle streams, or in wide rivers, *buoys*, or narrow small-decked skiffs, are used to buoy up the cable, and prevent it from impeding the velocity of the bridge, by dragging

on the bottom. They are placed about one hundred and twenty feet apart, their number depending on the circumstances of the case. The cable should be borne on small supports slightly elevated above their decks.

370. Flying-bridges of boats, or rafts, abandoned solely to the action of the current, may be found very serviceable in effecting a passage by surprise; the bridge is cast loose from the shore and directed on the opposite one, by means of an oar astern, to keep it in the proper position to receive the most favorable action of the current. The platform in this case may be surrounded by an enclosure of timber, or bags of wool or cotton, so as to be proof against the enemy's musketry and case-shot in case of an alarm.

371. *Trestle Bridge.* The form of a *trestle* is the same as what is termed a *carpenter-horse,* that is a horizontal beam supported by four legs.

372. The horizontal beam, termed the cap or *ridge-beam,* is usually of eight-inch scantling, and from twelve to sixteen feet long. The legs

are of four-and-a-half-inch scantling; they have a spread towards the bottom, the distance between them across being equal to half the height, and lengthwise of the cap, their inclination is one-twelfth of the height; they are fastened to the cap, about eighteen inches from the ends, by nails, the side of the cap and the top of the leg being properly prepared for a strong, accurate fit. The legs are connected either in pairs, or else all four by horizontal pieces of three-inch scantling; sometimes diagonal pieces, going from the top of one leg to the bottom of the opposite one, are used.

373. Bridges on trestles are principally useful in crossing small streams not more than six feet deep. They also serve to connect a boat bridge with the shore, in shallow water; and are used as the supports for the bridges of communication across the wet and dry ditches of field works. They present the advantage of being formed of materials which can always be procured at hand, either by cutting down forest trees, or by demolishing the floors and partitions of houses. More-

over, when constructed of light wood, they may be taken to pieces, and be easily transported on horses, or mules, in a wooded or broken country.

374. The trestles should not be placed farther apart than sixteen feet between the ridge beams, for balks of the ordinary dimensions; the balks should jut at least one foot beyond the ridge beams. The flooring, fastenings, &c., are the same as in a boat bridge.

375. When the bed of the river is of soft mud, or is hard and uneven, it is difficult to place the ridge beams horizontal; in the former case to make the legs sink equally, they may be let into sills, which should jut a few feet beyond them.

376. The action of the current is counteracted, either by anchors, or else by attaching each trestle to two cables, stretched across the stream above and below the bridge; stout stakes, or piles, might also be used for the same purpose. Another plan consists in making an open *watling*, or a net work, of tough twigs, or cords,

around the legs, near the bottom, and filling it in with broken stone.

377. The best manner of placing the trestles in deep water is by means of a raft, or a boat, so prepared that the trestle can be readily lowered into the water over the side.

378. *Temporary bridges* are readily constructed in wooded countries. The officer, charged with establishing communications of this character, will always do well to imitate the ordinary constructions of the country, particularly if he has to employ the country people to form them. In swamps and marshy grounds, a road-way of timber or stout brushwood, or even of trusses of hay or straw, will frequently have to be resorted to. After opening a way of sufficient width, and cutting down the stumps, so as not to impede the passage of vehicles on wheels, the trunks of the largest trees are cut into lengths of about fifteen feet, and are laid across the line, at distances of ten or twenty feet apart; notches are cut into the cross-pieces, about eighteen inches from each end, and sufficiently large to receive

trunks of trees from nine to twelve inches in diameter; these trunks should be cut into pieces of as great length as the tree will afford. Across these longitudinal pieces the bodies of small *saplings*, from three to four inches in diameter, and fourteen feet long, are laid, and small brushwood, earth, or straw, may be laid over them, to make the road-way more regular. In some cases the road-way may be formed by laying the saplings without any other substructure.

379. *Fords*. Infantry can ford a stream when the depth of water is not greater than three or four feet; cavalry when the depth is not more than four feet; and artillery, a depth of two-and-a-half feet. In cases of great emergency, particularly in gentle currents, infantry and cavalry may pass fords of greater depths than those just laid down; but safety, and a regard to keeping the ammunition in a good state, prescribe the depths here given.

380. The bottom of a ford should be of solid gravel; if of mud, fine sand, or obstructed with large fragments of stones, the difficulty of pass-

ing carriages and large bodies of troops will, generally, be found insurmountable

381. Fords are frequently met with near elbows, and usually in an oblique position to the thread of the stream. In making a reconnaissance for the purpose of seeking for a ford, a small boat should be allowed to float down with the current, and soundings be made.

382. To prevent accidents whilst the troops are crossing, stakes should be placed in the direction of the ford, showing its width; ropes are fastened to these stakes, on a level with the water, to prevent loss of life. If the ford is crossed at night, lighted torches are tied to the stakes.

383. If a large body of troops should cross, the infantry should pass first, the artillery next, and the cavalry last, so that the bottom may not be cut up by the feet of the horses.

384. *Ice.* The heaviest guns may cross on ice which is six inches thick, and rests on the water under it; but, as a measure of prudence, plank should be laid under the wheels to distribute

their pressure over a greater surface, and also to counteract the effect of shocks.

385. When the ice is but three inches thick, infantry and light guns can cross it with safety; but the men should be sent across in small detachments, and plank, or straw, should be placed under the wheels of the guns and the feet of the horses. By sprinkling water over the straw, it will, by its freezing, form a more compact roadway. Heavy guns must be carried over on *sleds*, taking every precaution to distribute the weight of the load over a large surface.

386. Ice which does not rest on water cannot be trusted to, unless it is at least six inches thick, and even then it should be crossed with great caution.

387. An army should never trust either to fords, or to ice, for keeping open its communications; for a freshet, or a thaw, would place it in the most critical state.

388. *Preservation and Destruction of Bridges.* Floating bridges are liable to injury from oscillations, caused by high winds, and the crossing

of troops; from sudden freshets; and from the shock of floating bodies.

389. Besides the usual means which are taken to insure solidity, by preventing oscillations, the manner of conducting the passage of troops requires great attention. Infantry should pass in half battalions, with an interval of forty or fifty paces between each half battalion, and in order of route, by a flank. Cavalry should dismount, and lead their horses, leaving forty or fifty paces between each squadron. Carriages should not cross at the same time as troops, and in no case should they be suffered to pass each other; the intervals between them will depend on the weight of each. Finally, the pontooners should be at their posts, to keep the cables and spring-lines well stretched, and to tighten the rack-lashings.

390. A firm anchorage, and a solid construction, are the best securities against the action of high winds, and the sudden rise of the river. To prevent the effects of the latter, the pontooners ease off the cables as the river rises; and

when danger is apprehended for the safety of the bridge, it must be cast loose, and suffered to swing round to one of the banks, every possible precaution being taken to prevent its gaining a great velocity during the movement.

391. *Pumps* and *scoops* should be provided to bail out in case of leaking or the breaking in of the waves.

392. In shallow streams, a stoccade may be formed, either obliquely across the river, or else with an angle up-stream, at four hundred or five hundred yards above the bridge; strong cables, or chains, are attached to the stoccade, on a level with the water, to arrest heavy floating bodies, which might injure the bridge by striking against it. In deep water, a floating chevaux-de-frise, made of heavy timber formed into a frame work of several feet in thickness, is thrown across the river at the same distance as the stoccade above the bridge. The chevaux-de-frise is confined by anchors and cables, and also by large piles, driven into the banks, to which it is firmly attached.

393. Posts of observation are placed to watch

the stoccade, or the chevaux-de-frise. They are provided with boats, ropes with grapnels, and a few pieces of artillery, for the purpose of arresting all floating bodies.

394. If a flying-bridge is placed, in connection with the boat-bridge, it should always be below it; so that no accident may occur from the flying-bridge parting from its cable.

395. To destroy a bridge of the enemy, heavy floating bodies, *fire-boats*, and *infernal machines*, are resorted to.

396. When the object is to destroy the bridge by a shock, a raft is formed of several large trunks of trees. A short and strong *mast* is solidly attached near the head of the raft, for the purpose of carrying away the flooring, if the raft should pass between the intervals of the boats. Large boats, heavily laden with stone, may be used for the same ends; the object in both cases being to acquire sufficient momentum to destroy the bridge.

397. Fire-boats, loaded with combustible matter, are sent down with the stream to burn the

bridge. Hollow projectiles, heavily charged with powder, are placed in the fire-boats to deter, by successive explosions, the pontooners from approaching them. Infernal machines are arranged so as to explode by the shock of the boat against the bridge.

In an attempt to destroy a bridge by these means, a number of them should be sent down at once, to increase the chances of success. They are conducted by boatmen as near to the bridge as they can approach with safety, and are then abandoned to the action of the current.

398. When a bridge is no longer of service, it may be destroyed, to prevent its being of use to the enemy, either by sinking it, burning it, or blowing it up. Combustible matter and cannon will serve for the two first mentioned methods, and a barrel or two of powder, placed below the flooring, and fired by means of a powder-hose, or a slow-match, will answer for the third.

399. A stone-bridge can be destroyed by making a trench, eighteen inches deep, across the roadway, over the crown of an arch; placing three

hundred, or four hundred pounds of powder in it, and covering it over with plank, and a heavy weight of earth, and exploding it by means just mentioned. Or, if pressed for time, several barrels of powder exploded at the same time in the open air, over the crown, will effect the same purpose.

400. *Passage of Rivers.* There is no military operation of a more delicate character, and of more doubtful success, than the passage of a river in the face of the enemy, whether acting offensively, or in retreat.

401. A passage by *main force* must always be accompanied by the most bloody results, if the enemy displays even ordinary courage and good conduct; and it cannot be attempted, with the slightest prospect of success, unless the troops are covered by a very superior artillery, posted so as to prevent the enemy from charging the troops in the act of disembarking, before they can form in order of battle, by crossing its fire in advance of the point where the passage is to be effected. The main difficulty in this operation

consists in getting the troops, first landed, to stand firm until, by successive reinforcements, they find themselves in sufficient strength to advance on the enemy. And the best plan to gain this indispensable time, is to throw over some squadrons of cavalry among the first, which, by successive charges, may keep the enemy at a distance.

402. In attempting a passage, by main force, a point must be selected on the opposite shore where the banks are not steep, wooded, or broken, and where the water is sufficiently deep to float the boats to a firm landing, in order that the troops may be able to form readily in order of battle, and to push forward without danger from a surprise.

403. When the opposite shore is occupied by the enemy in force, and is vigilantly guarded, there remains no way to effect a passage but by *stratagem*. To effect this, the bridge equipage is transported, secretly, to some point near that where the passage is to be made. If there are affluents near this point, or islands, they may be

taken advantage of to launch the boats undiscovered, and even to construct the bridge in parts which are afterwards floated to their position and put together. Demonstrations are made on several points by the army, to call off the enemy's attention from the real point, and troops are thrown over, after night-fall, in row-boats, and on flying-bridges; and the bridge is immediately commenced, and should be in readiness for the passage of the army by early dawn.

404. To protect the operation from a surprise, the troops first thrown over should occupy all the most accessible approaches, within twelve hundred or fifteen hundred paces, in advance of the point of passage; and other troops should throw up a slight intrenchment, like the *parallel* used in a siege, about two hundred paces in advance of the ground to be occupied by the tête-de-pont. The object of these works is to keep the enemy off, should he present himself in force; and to gain time for all the army to pass over, and form, without confusion, in order of battle under cover of these first dispositions.

405. A retreat across a river is necessarily made in the face of the enemy; and here, more than in any other position, an army is called on to show the coolest and most determined courage, for its safety depends on the strictest observance of good order. After the bridges are ready, and the tête-de-pont is in a defensible state, the movement of retreat commences, under cover of batteries of heavy guns on the opposite shore, these batteries having been thrown up at the earliest opportunity, and placed to take the enemy in flank.

406. The garrison of the tête-de-pont forms the rear guard, and retires last, leaving a few companies of grenadiers in the stoccade works, at the head of each bridge, for the protection of the pontooners, whilst they are preparing the bridge to be swung round.

In a retreat, a single bridge should never alone be relied on, as the slightest accident happening to it might compromise the safety of the whole army.

CHAPTER XIII.

MILITARY RECONNAISSANCES.

407. By the term *Military Reconnaissance*, is understood a detailed examination of any extent of country, whether large or small, with a view of ascertaining its resources for the movements and subsistence of troops. This examination may be of a *special nature*, as, for example, a proposed line of march; the character of a defile, river, or an enemy's position, &c.; or it may have a more general bearing, as the collecting of information for an entire campaign, or the exploring of a large district of country.

408. The importance of accurate reconnaissances, of whatever nature they may be, cannot be too strongly insisted upon; for all military operations must be based upon them, and any oversight, or neglect, in this respect, may involve

the safety of an army, and through it the destinies of a nation.

409. When the reconnaissance is of a special character, the officer charged with it should confine his operations strictly within his instructions, and not waste time upon any object, which might with propriety be brought into view in a more general reconnaissance; but when the object is to obtain information for the operations of a campaign, or for distant expeditions nothing should be overlooked which may bear, even in a remote degree, on the movements and subsistence of an army.

410. A reconnaissance consists of two parts: a *map*, and a *descriptive memoir.*

411. The map should exhibit the topographical features of the country; and the memoir should supply whatever features cannot be shown on a map, and also furnish information on the resources of the country for the subsistence and movements of troops.

412. The style of the memoir should be concise and clear; and nothing should be contained

in it which is not strictly relevant to the object in view. The information furnished by it should be, first those topographical features which a map cannot so well exhibit, as the nature of the face of the country; the quality of the soil; the height and declivities of mountains; on what side they are accessible, and for what kind of troops, and the means of transportation over them; the character of forests, water courses and other obstacles. Second, the lines of communication, as roads, rail-roads, canals, navigable streams, and all the circumstances in connection with the subject of transportation; the advantages and defects of military positions for the offensive or defensive. Third, every thing which relates to the culture and produce of the country; its fertility or barrenness; its resources in provisions and forage; the size and population of the cities, towns, villages, &c. Fourth, the commercial and manufacturing resources of the country; the number of mechanics of the different trades that are most necessary for the supplies of troops.

413. Such is an outline of the operations of

reconnaissances in general; and from it may be gathered the acquirements which the officer, charged with making them, is supposed to possess. To a familiar acquaintance with the duties of the topographical engineer, he should unite a large fund of general information on statistics and the natural sciences; a fund which nothing but a life of daily study and observation can supply.

414. The reconnaissances required by the daily operations of a campaign, are usually of a special character; such as examining a line of march or a military position. A special reconnaissance consists, for the most part, of a sketch in pencil, made with all the accuracy that the means at the disposal of the officer admit of; and a memoir.

415. Before commencing the reconnaissance; all the previous information, which can be obtained from maps, and published descriptions of the country, should be carefully collected, and afterwards compared with the officers operations; this will serve, not only to abridge the labor, but also to give greater or less confidence in the pub-

lished information on parts which cannot be examined for want of time, or from other causes.

416. In examining the face of the country, *wood-cutters*, *hunters*, *trappers* and *Indians*, are the best persons to apply to for information, and they should be chosen for guides. To obtain information on the resources of the country, *magistrates*, *clergymen*, *inn-keepers*, *tax-gatherers* and *farmers*, should be questioned; but from whatever source the information may be derived, it should be received with due caution, and only be acted on when confirmed by strict cross-examination of all the parties from whom it comes.

417. The nature of a special reconnaissance has already been shown in the two preceding Chapters, and but little more need therefore be added under this head; particularly, as special instructions, arranged under a tabular form, should always be supplied to the officer, charged with the operation, both as a guide to himself, and to enable the officer to whom the report is made to see at a glance the exact state of the case.

418. The reconnaissance of the enemy's posi-

tion is one of the most important of a special nature. The officer charged with this service should be thoroughly skilled in the duties of *out-posts;* as his operations must, in most cases, be carried on under the protection of a detachment, subject to his orders. In performing this service, he should never lose sight of the sole end in view, *the obtaining of information on a particular point*, and all his dispositions should be made to attain this end. For this purpose he must vigilantly guard against a surprise, and avoid an action with the enemy, unless the safety of himself and his detachment shall render it necessary; for the procuring of the required intelligence, which is to regulate the movements of the entire army, is paramount to every other consideration.

419. In reconnoitering the enemy's position, the extent of the ground which he occupies, and its character, should be first ascertained; the character of the approaches to his position, and the disposable means that can be found at hand to take advantage of the approaches;—the disposition

of his troops, and the arrangement of his defences ;—the towns, villages, and other unoccupied sites along his line which might be occupied with advantage by either party ;—the character of his line of retreat, and the defensible positions in his rear ;—the possibility of turning his position, and cutting his line of operations. In conducting this examination, the officer, of course, will be called upon to make other special reconnaissances of a more minute character; on the nature of the roads, by-paths, bridges, defiles, rivers, fords, marshes, swamps, forests, &c.; in which will be minutely stated the facilities they may severally afford, and the obstacles they present, to the movements of troops of all arms. The form of these minute examinations must be furnished the officer; for no presence of mind, and tenacity of memory, are sufficient to call to mind all that is necessary to be done in an operation which, from its very nature, must be of a hurried character.

420. A reconnaissance of a position to be defended temporarily, or to be occupied as an in-

trenched camp, is made with the same care as one for offensive operations. It should state, not only the advantages and defects of the position itself, but should point out the character of other positions in its rear, that might be occupied to check the enemy's progress, should he force the first.

421. The *Military Sketch* of a reconnaissance should be made with all the accuracy that the means at the officer's disposal will admit of. Surveying instruments are not always to be found in an army, and their places must be supplied by the officer's own ingenuity; distances are ascertained approximately, either by pacing them off with a uniform step, and estimating them by the length of the pace, or by the time required. The position of objects may be laid down by the intersections of lines, or by estimating in a rough manner the angles between different points. Slopes and declivities may be judged of by the greater or less time which it may take a man, or a horse whose action is well ascertained, to ascend or descend them.

422. A *pocket sextant*, a small *protractor*, and an *ivory scale*, are invaluable instruments for this service. They not only furnish the means of greater accuracy, but they essentially abridge the labor and time required with less complete means.

CHAPTER XIV.

OUTLINES OF THE GENERAL PROPERTIES OF PERMANENT WORKS AND THE METHOD OF ATTACKING THEM.

423. The term *Permanent Fortification* belongs to that branch of the *Art of Fortification* where means of a durable character are used to strengthen a position. Permanent differs from Temporary Fortification, not only in the character of the means used, but also in offering a more formidable obstacle to the enemy from the greater strength of its profile.

424. This difference occasions a great disparity, both in the measures and the time required for the attack of permanent and field works. In the latter, a hasty assault decides the fate of the affair; in the former, the enemy is obliged to provide vast resources both of men and materials,

and to approach the work attacked slowly, and with the greatest caution, under the cover of intrenchments, by which his troops are sheltered from the fire of the work, and he is enabled leisurely to destroy its defences.

425. Permanent works may be divided into two general classes, *Fortresses* and *Forts.* The term fortress is applied to fortified towns alone, and the term fort, to a work containing only a garrison.

426. The character of the fortification is the same in both classes; consisting, in its most simple form, of an elevated and wide mound of earth, termed the *rampart*, which encloses the space fortified, of an ordinary parapet surmounting the rampart, and of a wide and deep ditch which surrounds the whole.

427. These parts of the profile serve the same purposes as the corresponding parts in the profile of a field work; the most striking difference between the two consists in the rampart, which, from its height, gives a very commanding posi-

tion to the parapet, and greatly increases the obstacle presented to the enemy.

428. To give both strength and durability, the scarp and counterscarp are revetted with walls of masonry which sustains the pressure of the earth, protect it from the effects of the weather, and, by their height and steepness, presents an insurmountable obstacle to an assault by storm.

429. A fortification thus constituted would be sufficient for the protection of troops within it; but would not admit of exterior operations, because it affords no shelter beyond the ditch. Therefore, to procure the facility of manœuvering on the exterior, a low work, in the form of a glacis, is thrown up a few yards in front of the ditch, and completly enveloping it. The space between this work and the ditch is termed the *covered-way*, because it is covered from the enemy's view.

430. The simplest form then of an effective profile for permanent fortification, consists of a covered-way; a wide and deep ditch, with a scarp and counterscarp of masonry; and a ram-

part, which, from its height and width, will give a commanding position to the parapet, and sufficient room behind the parapet for the necessary manœuvres of the troops whilst in action.

431. The general principles of permanent fortification, as now practiced, are the results of centuries of experience in the attack and defence. The problem presented for the solution of the engineer consists in making such a disposition of his works that no point, within the range of their cannon, shall afford a shelter to the enemy; that they shall enclose the greatest space with the smallest perimeter, without sacrificing the reciprocal protection of the parts, afforded by a flanking arrangement within the medium range of arms; that no defensive dispositions, which can be destroyed by the enemy's distant batteries, shall be exposed to their fire; and finally, that the works shall be secure from an attack by storm.

432. To satisfy these conditions, it was found, that the space to be occupied must necessarily be enclosed by a series of bastions connected by

curtains; that the line of fortification must be continuous, and consist of a wide and deep ditch, and a high and steep scarp of masonry, to be perfectly secure from an escalade; and that the masonry of the scarp, which is the only part that can be destroyed by a distant fire, must be covered from this fire by the glacis of the work which forms the covered-way.

433. From the range of the fire-arms that are used in the defence, it was found that the distance between the salients of the bastions should not exceed four hundred yards; and that for a reciprocal flanking arrangement, the length of the curtains should not be less than twelve times the absolute relief.

434. To secure the work from escalade, experience has fully proved that the scarp wall should not be less than thirty feet high, and that the top of it should not be above the crest of the glacis.

435. With regard to the dimensions of the rampart, they must depend on those of the ditch, and a suitable regard to that economy by which all human efforts are circumscribed. The width

of the rampart required for the free manœuvres of troops has been fixed at about forty feet; and its height should give the parapet a command of at least twenty feet over the exterior ground.

436. The dimensions of the parapet are the same as those for the profile of field works of the strongest class.

437. The fortification by which the space fortified is immediately enveloped, is termed the *Body of the Place*, or the *Enceinte*. It is seldom that a permanent work consists simply of an enceinte, with its ditch and covered-way, particularly if some of its points are, from their locality, weaker than the rest. Other works are, usually, added to strengthen these weak points; they are termed *Out-Works* when they are enveloped by the covered-way, and *Detached* or *Advanced-Works*, when placed beyond it.

438. The object of these works is to lengthen the defence, by forcing the enemy to gain possession of them before he is able to make a breach in the enceinte.

439. The principal out-work is one in the form

of a redan, termed the *Demi-Lune*, which is placed in front of the curtain. This work adds to the main defence by a cross fire on the bastion salients, which are the weak points of the enceinte, and when there are demi-lunes on adjacent curtains, the bastions between them are placed in strong re-enterings, thereby forcing the enemy to gain possession of the demi-lunes before he can penetrate, without great labor and loss of life, into these re-enterings. The main entrances to the work are usually through the curtains, which being the most retired parts are also the most secure; the demi-lunes also serve to cover these entrances, and to guard them from a surprise. This, at first, was the only use to which the demi-lune was applied; but engineers soon discovered that, by enlarging it, new properties were developed which at length caused it to be regarded as an indispensable accessory in the defence of weak points.

440. The demi-lune, when of suitable dimensions, has also several other important properties, upon which it does not come within the scope of

this chapter to dwell. Its ditch is sometimes on the same level with the main ditch; sometimes it is higher; but in all cases the communications between the two, and also with the demi-lune itself, are arranged so as to be easy and secure.

441. Situated between the two flanks of the bastions, and directly in front of the curtain, a small low work, termed the *Tenaille*, serves to mask the scarp wall of the curtain and flanks from the enemy's batteries along the crest of glacis. This mask is of very great importance, since, by preventing the enemy from making a breach in either the flanks or curtain, it will force him to make it in the face of the bastion; the flanks will thus be preserved for the defence of the breach, and the enemy will not be able to turn the temporary, or permanent works, which may be constructed within the bastion to prevent him from gaining possession of the main work, by an assault of the breach, which he would be able to do could he effect a breach at the same time in the curtain or flanks.

442. The covered-ways of the bastion and

demi-lune, form a strong re-entering at their point of junction, of which advantage is taken to arrange a small redan whose faces flank the glacis of the two covered ways. The space enclosed by this work, which is a part of the covered way itself, is termed the *Re-entering Place of Arms.*

443. The parts of the covered ways in front of the salients of the bastion and demi-lune, are termed the *Salient Places of Arms.*

444. The places of arms are so called, because they serve for the assemblage of bodies of troops who are to act on the exterior.

445. Small permanent works, termed *Redoubts*, are placed within the demi-lune, and the re-entering place of arms, for the purpose of strengthening those works.

446. It is a received military principle, that the garrison of a work is no longer in safety, when it can be carried by storm, unless they are provided with a secure point of retreat. It is to effect this purpose that redoubts are constructed. The one in the re-entering place of arms secures the cover-

ed ways from an attack by storm; and that in the demi-lune forces the enemy to advance gradually, and with the greatest precaution, to gain possession of the breach in the demi-lune, and being provided with flanks, which, from their position, have a reverse fire on the breach in the bastion face, the enemy is forced to make himself master of it before he can venture to assault the breach in the bastion.

447. Works, termed *Interior Retrenchments*, which have the same properties as a redoubt, are constructed within the bastion. When the interior retrenchment is sufficiently elevated to command the exterior ground, it is termed a *Cavalier*.

448. The protection afforded by a redoubt to another work, is not by offering a place of safety into which the garrison of the work can retire when driven out of it; but in covering the retreat of the garrison by a warm fire, which will check the advance of the enemy, and enable it to retire behind the redoubt, and not into it; for such an operation would inevitably create confusion in

the redoubt, and might jeopard its safety, as the enemy, in a hot pursuit, might enter it pell-mell with the retreating troops.

449. The crest of the glacis is broken into an indented line for the purpose of obtaining a flank and cross fire on the ground in front of the places of arms.

450. Traverses are placed at intervals along the covered ways; they serve to intercept the projectiles which enfilade the covered ways, and also to defend them foot by foot; enabling the troops to retreat from one part of the covered way behind the traverse under the protection of its fire.

451. The communications of the works consist of arched under-ground outlets, termed *Posterns*, of *Ramps*, and *Stone Stairs*. These communications are placed in the most secure parts of the works, so that the ingress and egress may be without danger.

452. With regard to the relief of the outworks, as a general principle those most advanced should be commanded by those most re-

tired. This principle is applied in all the works, except the tenaille and the redoubt of the re-entering place of arms. The former must not mask the fire of the bastion flanks along the main ditch; and the latter must not mask the fire of the bastion faces upon the glacis of the demi-lune covered way. To satisfy these conditions, the two works must be commanded by the demi-lune, which is more advanced than either of them; but, by the process of defiladement, they are both so arranged that the enemy will not have a plunging fire into them from the demi-lune.

453. All the fortification, comprehended between the capitals of two adjacent bastions and the glacis, is termed a *Front of Fortification*, or simply a *Front*. It is taken as the unit in permanent fortification.

454. The usual method of effectually protecting any point, is by a flank fire; but, owing to the locality, or to some other cause, it may not be practicable to make a flanking arrangement; to supply its place, dispositions, termed *Counter-*

EXPLANATIONS OF PLATE XI.

Fig. 69. Shows a plan and longitudinal section of a Raft prepared for a bridge; A are the logs of the raft; B cross-pieces to confine the logs; C sills to support and raise the bridge.

Fig. 70. Shows a perspective elevation of a Flying Bridge. A, A are the Boats; B the Platform; C, C the Uprights; D the Cross Pieces; E the Sliding Pulley for the cable; F the Cable; S, S Buoys or Skiffs for the cable to rest on.

Fig. 71. Shows a perspective elevation of a Trestle; A is the Cap or Ridge Beam; B the Legs; C Cross-pieces to strengthen the legs.

Fig. 72. Shows a Plan of a Regularly Fortified Front.

A, A, . . A is the Enceinte, or Body of the Place.

B, B the Bastions.

C, C . . C the Main Ditch, or ditch of the enceinte.

D, D . . D the bastion and demi-lune Covered-Ways.

E, E the Re-entering Places of Arms.

F, F, F, the Salient Places of Arms.

G, the Demi-Lune.

H, H the demi-lune Ditch.

J, the demi-lune Redoubt.

L, L the Ditch of the demi-lune redoubt.

M, M the Redoubts of the re-entering places of arms.

N, N, Ditches of the redoubts.

a, a . . a Traverses of the covered-way.

o the Tenaille.

Fig. A. Shows a Section of the enceinte, main ditch, and covered way of Fig. 72.

A is the Rampart; of which a b is the Slope, and b c the Terre-Plein.

B is the Parapet; of which c d e f g h is the outline.

C is the Main Ditch.

D the Scarp Wall.

E the Counter-Scarp Wall.

F the Embankment of the covered-way; of which m n is the terre-plein, n o p q the outline of the Banquette and Interior Slope, and q r the Glacis.

Plate. II

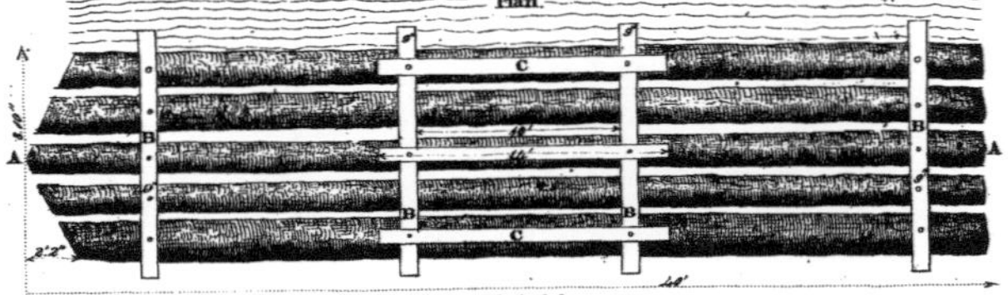

Plan.

Fig. 69

Longitudinal Section.

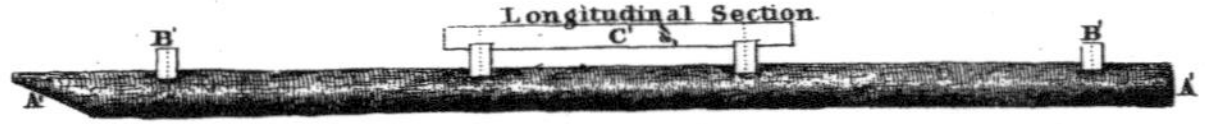

Fig. A
Section thro' the line m'n' Fig. 72.

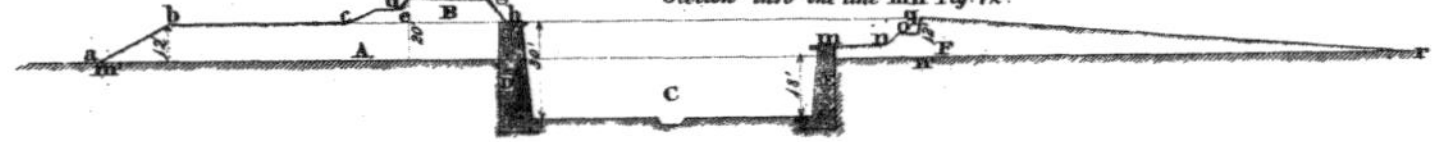

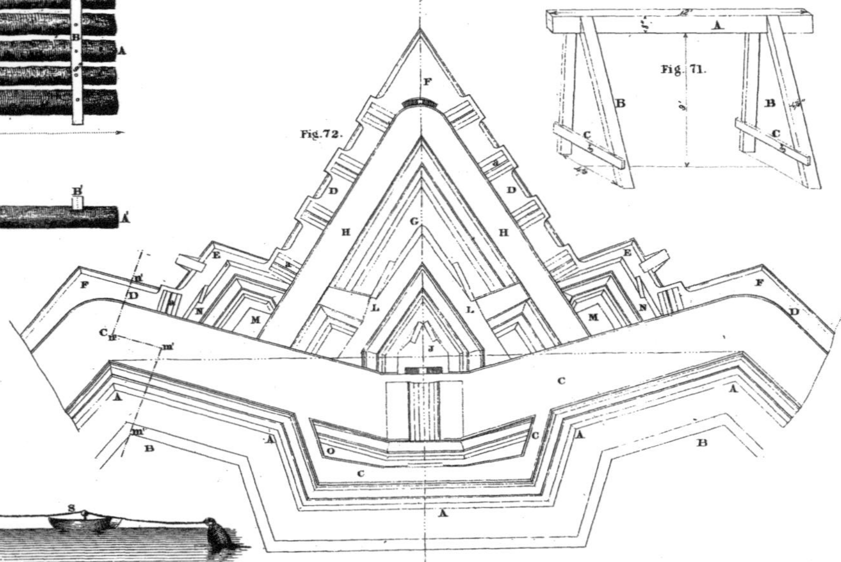

Fig. 71.

Fig. 72.

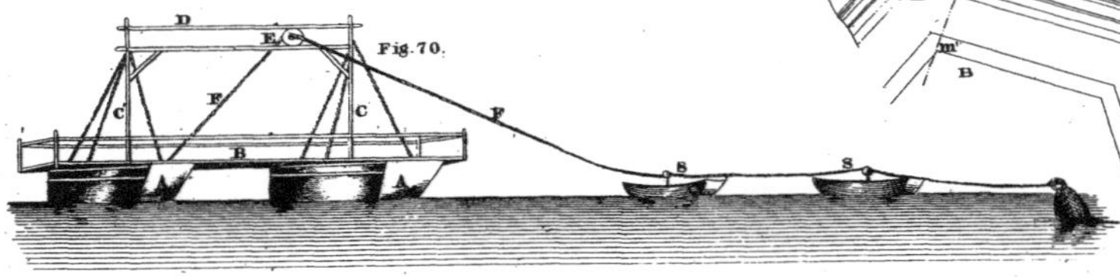

Fig. 70.

scarp Galleries, are made behind the counter-scarp, with loop-hole defences for a reverse fire. This arrangement approximates the nearest to the military solution of the problem *to see without being seen*, since from the position of these galleries, the enemy will not be able to bring his batteries to bear on them, whilst they will present a formidable impediment to all of his operations in the ditches.

455. For seacoast defences, embrasures are made through the scarp wall, and the artillery is protected from shells by an arched bomb-proof covering over head. This arrangement is termed a *Defensive Case-mate.* This method of defence is only efficacious against a sea attack; for on the land side, where the enemy can approach regularly, case-mates would be immediately destroyed by his batteries, and the loss of life would be far greater in them than in an open defence, owing to the fragments of stone which each shot striking an embrasure would cause.

456. Case-mates are also used simply as bomb-proof shelters for the troops and materiel; and

no small fort indeed which can be bombarded is tenable without them.

457. Attack. The measures for the defence resorted to in permanent works, are of a character to resist, with a certainty of success, multiplied attacks, with the most powerful means, if made openly. This certainty forces the enemy to provide a large number of troops and materiel; and to approach the work slowly, and always under cover; making himself master of the exterior ground foot by foot, and proceeding forward only after having made such a disposition of his troops and works in his rear as to repel all the efforts of the garrison to delay his progress. This method is termed the *Attack by Regular Approaches.*

458. A *Siege* is the operation of cutting off all communication with a work, and attacking by regular approaches. When the enemy confines his operation to a simple interruption of the communications, it is termed a *Blockade.*

459. The operations of a siege may therefore be classed under two heads; those which are

necessary to prevent all ingress to, or egress from, the work, and those which are required to gain possession of the work.

460. The first class of operations comprehend the *Investment*, and the establishment of the *Besieging Army* in intrenched camps around the work.

461. The investment is performed by detaching a strong corps in advance of the besieging army, who, by a sudden movement, surround the work, and seize upon all the avenues leading to it. The object of the investment is to cut off all communication with the work, and to secure every thing without it that might in any way prove of service to the garrison. To effect this, the enterprise should be conducted with despatch and secrecy; and the troops selected for it should be composed principally of cavalry and light troops, depending, however, on the character of the country; the number of troops being in proportion to the strength of the garrison, and the means that they can employ to oppose the investment.

462. During the day-time, the investing corps will keep beyond the range of the cannon of the works, occupying the most favorable points to prevent excursions from them. At night the detachments will close in upon the work, so as to form a continuous chain of sentinels around it. Engineer officers, who accompany the investing corps, will make a complete reconnaissance of the work, and the exterior ground, for the purpose of facilitating the ulterior operations.

463. In the mean time, the besieging army takes up its line of march, and arriving within the vicinity of the work, the different corps are distributed around it in camps, which are selected before hand, beyond the range of the garrison's cannon. The besieging army immediately proceeds to intrench itself, by throwing up works to prevent succours of troops, provisions, &c., being thrown into the place; and also other works to restrain the excursions of the garrison.

464. The first line of works, to prevent succours, is termed the *Line of Circumvallation;* the other is the *Line of Countervallation.* The

first is usually a continuous line, as it is most suitable for the end in view; the second consists of detached works to cover the weak points of the camp, and particularly the parks of the engineers and artillery, in which all the siege train of these two arms are collected. These lines are about six hundred yards apart to leave room for the camps.

465. The camps of the different corps are connected by removing all the obstacles between them, and opening communications for a free circulation of the troops, so that succour can be promptly despatched to any point which is in danger from an attack by a superior force.

466. Whilst the lines are constructing, the engineer officers are engaged in making a complete survey and plan of the work and its approaches; and in superintending the preparation of all the materials required for the construction of their own works, so that they may be carried on with vigor after they are begun.

467. The object of the survey and plan is to enable the general, superintending the siege, to

select the most suitable point to approach the work. This point is termed the *Point*, or *Front of Attack*. In selecting it, regard must be had to the strength of the point itself; the nature of the ground over which it must be approached; and the more or less ease with which the requisite *materiel* can be brought to this point.

468. The salients are usually the weakest points of a work, because they are easily enveloped by the enemy's approaches, and he will therefore be able to concentrate a heavy fire on them. Solid rock, a stony, or a marshy soil, present the greatest difficulties to the construction of the works of attack. And a good road, or navigable stream, affords the greatest facilities for the transportation of all the materiel required during the siege.

469. The object of the besiegers being to gain possession of the point of attack, with the least sacrifice on their own side, the whole of their measures should tend to the attainment of this end. To effect this, the troops must be able to approach the work under shelter; its fire must

be silenced; and a breach be made in its rampart.

470. The works under whose shelter the troops approach the work attacked, are termed *Trenches*, (Fig. 73,) of which there are two classes, the *Parallels* and the *Boyaux*. Each class consists of a ditch, or trench, of uniform depth, and a parapet of uniform height, formed of the earth taken from the trench, and thrown up towards the work attacked.

471. The parallel is a long line of trench, concentric with, or parallel to, the works of the point of attack, which it envelopes; and the boyaux are simply communications, in a zigzag direction, which lead to the parallel. The parallels and the boyaux both serve as a covered way, in which the troops can circulate with safety; but the boyaux are used exclusively for this purpose, whereas the parallels serve to contain troops who restrain the excursions of the garrison, and protect the workmen whilst carrying forward the approaches.

472. The fire of the work is silenced, by selecting suitable positions, near the parallels, from

which the faces of the work can be enfiladed, and its guns be dismounted by a ricochet fire.

473. The breach can be made when, the approaches having arrived at the crest of the glacis, the scarp walls are exposed to breach batteries, constructed along this crest, which firing through the ditches of the work, batter away the scarp to its very foot.

474. To lay out the trenches the alignement of the faces, and capitals, of the front of attack, as well as those of the collateral works, whose fire bears on the ground over which the approaches must be carried, are accurately marked out on the ground by stout pickets, and the distances from the most advanced salients to some point on their capitals are also determined.

475. The object of procuring these alignements is to mark the position and direction of the trenches and batteries. The boyaux are run in a zigzag direction across the capitals; because this is the shortest line to the salients of the work attacked which are the points first reached; and in this position they are less exposed to the

fire of the work, and are less in the way of the other works of the attack, than they would be if run on any other line.

476. The position of the boyaux and parallels is laid out on the ground with pickets, to which cords are attached, or simply with pickets alone. The position of the *First Parallel* is six hundred yards from the most advanced salients; it should embrace within its extent the faces of all the collateral works that protect the point of attack.

477. At the distance of six hundred yards, the fire of the work is not very troublesome; and it is within good range for a ricochet fire, if it be found necessary to erect the enfilading batteries near the first parallel.

478. Boyaux lead from this parallel to points in the rear of it, about fifteen hundred yards from the work attacked. At these points all the implements and materials of the engineers are collected, and they are therefore termed the *Depots of the Trenches.*

479. The approaches are begun by what is

termed the *Opening of the Trenches.* The workmen are assembled at the depots, are supplied with intrenching tools, and are marched and stationed along the line of boyaux, and that part of the first parallel which is to be first made. This is done just at the close of day; and troops, drawn up in order of battle, are stationed a short distance in advance of the workmen to protect them from the sorties of the garrison. The word is then passed along the line of workmen who commence opening the trenches, which, from this moment, are pushed forward without intermission until the work capitulates.

480. Whilst the first parallel is still in an incomplete state, boyaux are commenced to lead from it to gain the position of the *Second Parallel*, which is laid out parallel to the first, and rather less than three hundred yards from it; so that it may be still under the protection of the first, whilst in an incomplete state, should the garrison attempt a sortie against it. The object of this parallel is to protect the approaches as they are pushed forward upon the salients.

The same principle governs all the approaches: that is, they must be protected by a force superior to any that can be brought against them, and, in order that they may be succoured in time, this force must be nearer than the garrison to the workmen employed in the trenches.

481. The construction of the second parallel differs from that of the first, owing to the exposed state of the workmen to the grape shot of the work. To place the men under cover as speedily as possible, the line of the parallel is marked out, by placing gabions side by side, and when these gabions are filled, they afford a protection from grape shot.

482. With regard to the enfilading batteries, it is a disputed point whether they should be erected near the first or the second parallel. If the fire of the work attacked, is very warm and destructive, it may be necessary to place the batteries near the first parallel, to silence the fire of the work before the approaches can be pushed farther, but their effect, at this distance, is far less certain

than at the second parellel, where every shot can be made to tell.

483. Whatever may be the place of the batteries, their direction will be perpendicular to the face enfiladed. The batteries will contain cannon, howitzers, and mortars. The cannon destroy the guns and their traverses; the howitzers annoy the covered ways; and the mortars throw shells into any part of the work which cannot otherwise be reached.

484. The approaches are carried forward from the second parallel, so soon as the fire of the place is silenced; but here the musket soon comes into play, a weapon of a more certain aim than cannon, and therefore more destructive to the workmen. The approaches are now given over to the engineer troops, termed *Sappers*, who carry forward the trench, termed a *Sap*, slowly, but with constant progress day and night. The *sappers* advance the end of the trench, working on their knees, and shielding themselves in front from the enemy's fire, by a large gabion stuffed with fascines, or wool, termed a *Sap-Roller*,

which is rolled forward as they gradually advance.

485. When the boyaux are completed to within sixty yards of the most advanced salients of the work, the *Third Parallel* is commenced; the precaution having been taken, however, to place *Half Parallels*, between the second and third, to protect the approaches until the third is completed.

486. At the position of the third parallel, the besiegers have arrived on the immediate ground of the work attacked. Here the struggle begins to assume an attitude of equality. The attack loses its advantage of enveloping; it presents a front nearly the same as that of the defence; and shortly it becomes itself the enveloped; and then, if the resources of the garrison have not been entirely exhausted, each day will show but slight progress on the part of the besiegers, who are confined to their trenches, and exposed to all the disadvantages of a commanded and enveloped position.

487. After the completion of the third parallel,

measures are taken to drive the garrison from the covered ways, so that the breach batteries may be erected. This may be done in two ways, either by still proceeding by gradual approaches; or else an attack by storm may be made on the covered way, and whilst this action is going on, the sappers establish themselves along the crest of the glacis. The first method is slow, but certain, and accompanied with, comparatively speaking, but little loss of life. The second is rapid and more brilliant, but attended with great destruction of life, even in the most favorable cases, and is moreover seldom successful.

488. All open assaults of permanent works are attended with immense loss of life, and ought never to succeed where a garrison does its duty even with ordinary vigor. This is a maxim justified by the results of every attempt of this character on record; and it should never be lost sight of by the commanding officer of the siege. Nothing but a case of the most urgent necessity, where a day gained may decide the fate of the besiegers themselves, should cover from obloquy

any officer, whatever success may crown his order, who wantonly sheds the blood of brave men to attain an object, which a few hours delay would place within his reach, without perhaps the loss of a man.

489. To attack the covered way by storm, *Stone Mortar Batteries* are erected in front of the third parallel, for the purpose of throwing showers of stone into the covered ways. A detachment of chosen troops is stationed in the third parallel, to débouche from it, at a given signal, and rush into the covered ways. The sappers are also in readiness, with their implements, to make the required lodgment on the glacis. When every thing is prepared, a powerful fire is opened from the mortar batteries, and from all others, whose fire can reach the covered way, to drive the besieged from it, and from their other defences near it. This shower of projectiles ceases at a concerted signal, and the men composing the detachment rush from the parallel, and throw themselves into the covered way, driving every thing before them, and availing themselves of every

defensible point in order to prevent the besieged from sallying out, until the sappers, who follow close on them, have had time to construct a sap within six yards of the crest of the glacis, around the salient place of arms of the demi-lune covered way, into which they can retire for shelter from the covered way.

490. In the attack by gradual approaches, the sappers push forward the trench from the third parallel, under the protection of the fire from the stone mortar batteries, proceeding directly on the salient of the covered way, and when they have arrived at thirty yards from it, they extend the sap fifteen or twenty yards to the right and left, so as to embrace these salients, and throw up a mound of earth about ten feet high, from the top of which they can obtain a plunging fire into the covered way, and thus prevent the besieged from occupying it. This mound is termed the *Trench Cavalier.*

491. When the besieged are effectually driven from the covered way, the sappers push forward the sap to within six yards of the crest of the

glacis; and they make a trench entirely around the covered way, which is termed *Crowning the Covered Way.* After the covered way is crowned, breach batteries are erected around the salient places of arms, and from these positions breaches are made in the face of the demi-lune, and in the bastion faces.

492. The besiegers now begin their operations to get at the breach, which is done by forming an underground gallery of frame-work and boards, leading to the bottom of the ditch; this is termed the *Descent of the Ditch.* From the bottom of this descent a trench is made, by sap, across the ditch, and up the slope of the breach to its top, where a trench encircling the top of the breach is formed; these operations are termed the *Passage of the Ditch,* and the *Lodgment of the Breach.*

493. In some cases the breach is carried by assault, but the same objection is applicable here as in the attack of the covered way by storm. It should only be resorted to when the besieged

show, by their efforts, that the lodgment cannot be otherwise effected.

494. If the demi-lune is provided with a redoubt, this work must be taken before the bastion can be carried, because it has a reverse fire on the breach from its flanks. A breach battery will therefore have to be erected in the demi-lune against the redoubt, and when the breach is practicable, it may be assaulted at the same time as the bastion breach.

495. Similar operations will have to be made against the interior retrenchments, in the bastion. Finally, at the breach in the last work the garrison must capitulate, or undergo, according to the custom of war, the fate of those who are carried by storm.

496. Defence. The measures for the defence, although for the most part prepared before hand, will in some degree be subordinate to those of the attack. The fortifications are here supposed in a complete state, and supplied with all the necessary provisions and munitions of war for a full garrison.

EXPLANATIONS OF PLATE XII.

Fig. 73. Shows a plan of the Works of Attack of a Permanent Work.

A, A . . A the First Parallel.

B, B . . B the Second Parallel.

C, C . . C the Third Parallel.

D, D . . D the Boyaux leading from the Depots of the Trenches to the First Parallel.

E, E . . E the Boyaux from First to Second Parallel.

F, F . . F the Boyaux from Second to Third Parallel.

G, G the Demi-Parallels.

H, H . . H Enfilading Batteries on the First Parallel.

I, I . . I do. do. do. Second do.

J, J . . J Breach Batteries on the glacis.

O, O . . O Trench Cavaliers.

M, M Bastions of the Front of Attack.

N, Demi-Lune of the Front of Attack.

P, P Demi-Lunes of the Collateral Fronts.

Fig. B. Shows the General form of the Profile of the Trenches.

A is the Trench.

B is the Parapet.

C is a Berm, between the trench and parapet, that serves as a Banquette.

Fig. B.

Profile of the Parallels Fig. 73.

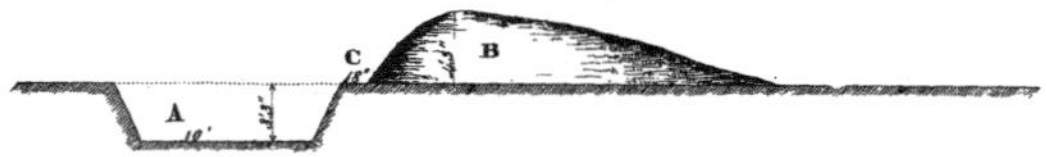

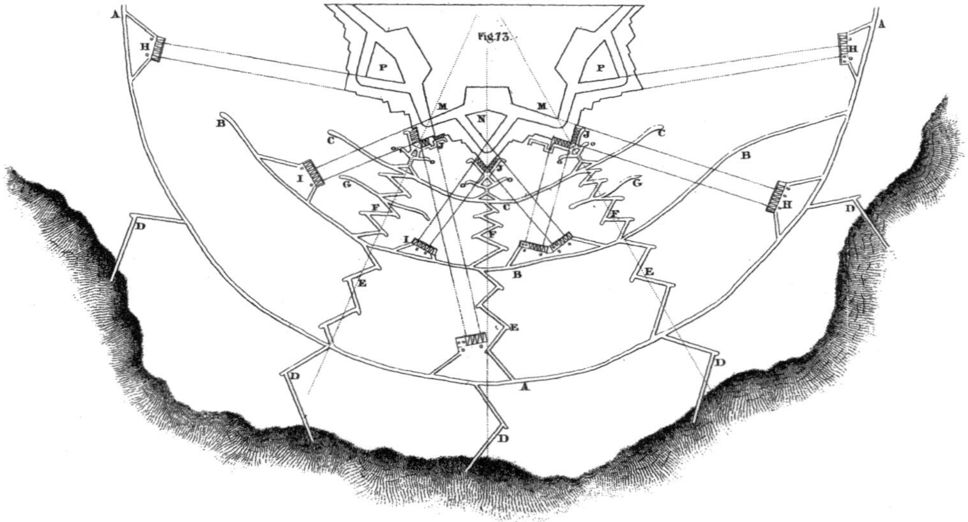

497. So soon as any indications of demonstrations against the work are seen, every thing on the exterior, within cannon range, which might afford a shelter to the enemy, must be removed. One gun at least, of long range, is placed in barbette on each salient of the enceinte and the demi-lunes; and at least two guns, loaded with case-shot, are placed in embrasure on each flank of the enceinte. These are simply measures of precaution to guard against a surprise, and to prevent reconnoitering bodies from approaching the work.

498. When the investment takes place, the garrison will resort to every expedient of open force and stratagem to cut off the enemy's reconnoitering parties; and, finally, when hemmed in by the superior forces of the besieging army, they will still watch every opportunity for annoying the enemy, by destroying his munitions, &c., without, however, jeoparding themselves, for they are now supposed beyond the reach of succours, and every life lost greatly increases the dispro-

portion between their resources and those of the enemy.

499. When all the preparations for opening the trenches are completed, the garrison will bring forward all its disposable artillery, and place it, in barbette, along the point of attack, this point having been ascertained by the movements of the enemy's troops, the establishment of his parks, depots, &c. Every night, about the presumed time for opening the trenches, fire-balls will be thrown out, to discover the enemy's movements; and when he commences his labors, a heavy fire will be opened on his laborers and troops, at the same moment from every point, and it will be kept up without intermission for some hours. This is the moment that a select detachment of the garrison will sally out from the collateral works, and at a given signal the fire will cease, to allow them to charge the enemy's flank. If this sortie succeeds, it should be vigorously followed up; the workmen are dispersed, and, during the confusion resulting from this, the detachment should, if there is the slightest prospect of suc-

cess, endeavour to penetrate to the enemy's parks, and destroy every thing in their way. This is the only moment that a vigorous sortie can be made with any success. The parallels once established, the odds become too great against the garrison to justify another attempt, except great remissness is observed on the part of the enemy.

500. The guns are withdrawn from the barbettes after the first night, and placed in embrasure on the most suitable points to retard the approaches; they are also covered in flank by traverses. A fire, mostly of ricochet, is kept up on all the works of the enemy, at regular intervals, until he has advanced within the range of grape and musketry, when both of these projectiles are resorted to; the artillery being served with the greatest coolness, and the best marksmen with muskets or rifles, being stationed in the covered ways.

501. This system is persevered in, until the enemy gets on the glacis, or the immediate ground of the work. Then commences a sys-

tem of *petite-guerre*, in which all means are taken to retard the work; sorties of a few picked men are continually made, to bayonet the sappers, and by throwing bags of powder, with slow-matches attached to them, against the sap-rollers, to blow them and the sapper up.

502. Finally, the covered way is vigorously defended inch by inch, if attacked by storm, the troops retreating from it under the protection of the fire of the works in its rear.

503. The defence of the breaches will chiefly depend on the measures of the attack; if these are by gradual approaches, the garrison will make frequent sorties, from the adjacent works in small bodies on the passage across the ditch, whilst others attack the lodgment on the breach in front. But if the besiegers make their preparations to carry the breach by storm, the besieged should closely calculate the chances of opposing with the bayonet this attack. It is true that the military code, as generally received among nations, prescribes that the garrison should not accept terms of capitulation until they have sus-

tained, at least, three assaults of the breach. But neither duty, nor honor, requires a brave man to sacrifice himself, when, all prospect of succour or of retreat being shut out, he finds himself in position, face to face, with ten to one against him, and entirely at the mercy of his enemy in case of defeat.

504. A breach can only be defended with a certainty of success, and without jeoparding the lives of the entire garrison, when a redoubt or interior retrenchment is in its rear, under the cover of whose fire the troops, placed at the breach to watch the enemy's movements, and delay his progress, can retreat with safety. When this measure of security is provided, the troops at the breach should use every human means, short of attacking with the bayonet, to prevent the enemy from advancing, and they must not retire before there is an appearance of imminent danger to the point of safety, from the enemy entering it pell-mell with the retreating detachment.

505. In a general summary of the ordinary

march of the attack and defence, and the application of the principles of permanent fortification, no allusion has been made to accessory measures of defence, which often furnish both parties with the most powerful resources, as *mines*, the uses of *water*, &c.; since these measures are not of universal application, owing to their requiring suitable localities for their employment; and they would, moreover, require explanations and details, which none but the professional engineer would probably understand, and which would offer no interest to any other.

506. *Uses of permanent works.* The subject of national defence is one which presents itself with such strong claims to the careful consideration, not only of the statesman and soldier, but of every private citizen, who has the deepest interest at stake, in the settlement of so momentous a question as the safety of the State, that a few words seem called for on the part which permanent works play in the military organization of a frontier; and the relations subsisting be-

tween their action and that of other means of defence.

A thorough discussion of this subject would embrace too wide a field to be brought within the present scope of the writer; for to do full justice to it, he should be obliged to enter into the development of the principles of *strategy*, and those higher considerations of the military art, which have been, not unaptly, termed the *sublime of war*. But there are a few reflections of too striking a character to be suppressed, and which the events recorded, not only in the history of foreign nations, but in that of our own country, must bring forcibly home to the breast of every thinking person.

To defend the utility, the indispensable necessity of the military organization of a frontier with permanent works, were to cite the experience of every people, of every age, who have been exposed to the disasters of invasion, whether upon their land or maritime frontier. Poland, owing to the want of such an organization, fell a victim to the greatest political crime of modern times.

Spain had well nigh been ensnared in an insidious attempt against her existence as an independent nation, of as dark a dye as the dismemberment of Poland, when the heroic resistance of Saragossa, and a few other fortified places, gave her people time to recover from their first surprise, and enabled them to rally under the protection of their strong holds, and organize that system of determined resistance, that "war to the knife," which at length freed their soil from the pollution of a foreign master's footsteps. France saw her naval power annihilated, and her commercial marine literally swept from the the seas, by the loss of a few naval pitched battles; and whilst the fleets of her enemy rode in triumph at her very threshhold, slept on in unbroken security, under the guns of her sea-board defences. At a subsequent period, when, overwhelmed by the united forces of Europe, her hitherto victorious armies were driven within their own territory, it is the opinion of able military men, that a few strong holds on the flank of the enemy would have enabled her great Captain, who contended with his usual genius against such unequal odds, to have retrieved her fortunes, whilst that enemy rushed on to seize her open unprotected capitol, and afterwards trembled in its possession at the thought of a re-action in the people.

No intelligent person, it is presumed, could be found, in the present advanced stage of military science, to advocate a system of defence with active forces alone, whether land or naval. The perishable nature of such armaments, the casualties to which they are exposed, the uncertainty of the issue of pitched battles, all point to some other means of defence, which, by prolonging the contest on one or more points, will increase the chances of retrieving any mishaps. To defend an open frontier of any nature, withoutthe aid of permanent works, requires the display of a large moveable military force, whose ultimate expense, long experience has but too amply proved, increases in a rapid proportion with the undisciplined character of the troops. This is the opinion of military men of all nations; and the testimony of our generals, whether of the militia, or of the regular army, goes to show, that whatever reliance may be placed on the valor and patriotism of our militia in times of emergency, still it is necessarily the most expensive force that can be brought into the field against an enemy, owing to the losses which are necessarily incurred by the employment of men totally unacquainted with the economy of a regularly organized army. This being the case, and the nature of our political institutions being such as to render a resort to a large standing-army dangerous to their existence, it behooves us to devise some means by which the smallest regular force may be called for in the defence of the State, and in which the militia may be employed, not only in the most efficient manner for the defence, but also most economically.

The best means, it seems to the writer, by which this can be effected, consists in a judicious organization of the frontier, with permanent works of such strength as each particular site may require. The defence of these works may be safely left to garrisons principally of militia; and as this arm of our national defence is composed of the most valuable members of society, this kind of service will be most appropriately intrusted to it, as it is less onerous to the individual than that of the field, and is moreover, attended with less destruction of life, both from the diseases incident to the exposure of the camp and the casualties of pitched battles.

As to the class of permanent works, which would be required for an efficient organization of our frontier, the writer does not feel himself called on to give an opinion, as this is a point which ought to be settled alone by a board of competent engineer officers. Without permanent defences, both our land and sea-board frontier would be left open, not only to the temporary inroads of an enemy's armies and fleets, but, as was seen during our last war with England, positions could be selected, and permanently occupied by him during the war, from which our entire coast might be kept in a state of continued alarm, whilst great detriment might be caused to our coasting trade, and to the line of the sea-board, from occasional marauding parties and small cruisers. With permanent works of a proper character at suitable points, the most of these evils would be obviated; our harbors would be secured, and our large road-steads, with even the extensive estuaries for which our coast is remarkable, could be effectually closed against an enemy's squadrons. These strong holds would serve as places of resort for our own vessels of war, when in less strength than those of the enemy, where they could lie in security, and watch for a favorable moment of action. They would s rve as covering points for our fleets, in event of any disaster happening, either from battle or from tempests. They would form the natural rallying points for floating batteries, moved by steam, by means of which the enemy would at all times be kept on the alert, and his position at any point far beyond the range of the guns of the work, would be rendered insecure, and in most cases, untenable, except at a great risk. In our land operations, these fortified points would play any thing but a passive part, if properly combined with moveable forces. In the hands of a skilful general, their part, on the contrary, would be eminently active. They might be made to cover, not only their own garrison, but large bodies of troops which might take up an unassailable position under their guns, and be in readiness at all times, to oppose the enemy in front, if of equal force, or to attack his flank or rear, if he attempted to treat them with neglect. They might be made to serve as points of safety, upon which the garrisons of weaker works could retreat, in case of necessity, or from which they could obtain succour if strongly pressed by a superior force. It is in this way that permanent works become truly formidable to an enemy, and enable us to guard distant points beyond the range of their cannon, without placing a man, or a gun, for the immediate defence of those points.

But the expense of such an organization is objected to, and this objection is indeed weighty, when it is reduced to a certainty, that not thousands alone, but millions must be expended before the state can be placed in this attitude of permanent safety. Can any other better, or more certain means, however, be devised, to effect the proposed end? Have those who would leave the tranquillity of the state to the care of moveable forces alone compared the expense of their plan, simply under a pecuniary point of view, with that of permanent works? Setting aside this cost, have they estimated the inevitable losses that must be sustained by agriculture, commerce, and all the useful and agreeable arts, not only from depredations by the enemy, but from the withdrawal of so many of the members of these professions from their daily avocations? Have they estimated that greatest of all misfortunes, arising from the loss of life of valuable members of society, which brings in its train the present misery, and future impoverishment of whole families, when deprived of their head? But time, it may be urged, will repair all these losses; a new generation will spring up in the peace procured by the blood of their gallant ancestors, who will restore the former prosperity of the country.

Time, it is true, may effect this. Time we know has done more. But are the sacrifices thus made, to be counted for nothing? Can they be weighed in a balance against any assignable amount of gold? But granted that to-day ought always to be made a willing sacrifice for to-morrow, are there no other interests jeoparded in a state of things, which would necessarily convert the whole country into one vast camp? Will time heal the wounds inflicted upon morals generally, during a period of such comparative disorganization? Will it recover public opinion from the consequences of such a disorganization, and restore to it that healthy tone so essential to a nation's true greatness and welfare? There may be losses—history tells us there have been such—which even time has not repaired, but by such throes as shake the foundations of civilized society, and utterly overthrow the works and institutions of man; and it is against the recurrence of these, that every enlightened patriot, made wise by example, should with prudent foresight guard, by a wholesome use of the means at his disposal.

NOTE A.

Manner of determining the proportion between the terre-plein of a Square Redoubt, the development of its interior crest, and the size of its garrison.

Fig. A.

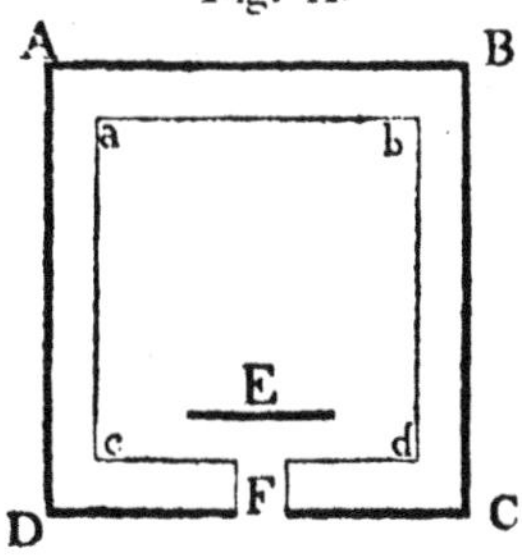

Let A B C D (Fig. A.) be the line of the interior crest; a b c d that of the foot of the banquette slope; F the outlet through the parapet; and E the interior traverse.

Representing by x, the number of yards in the side A B, the side a b will be equal to x diminished by twice the distance between A B and a b, which distance may be taken on an average at 9 yards; the area of the square b c, or of the terre-plein, will then be represented by $(x - 9)^2$.

Let the following notation be adopted :

x, the side of the square A B.

y, the number of the garrison.

r, the reserve taken from the garrison.

n, the number of ranks on the banquette for the defence.

s, the area of the terre-plein occupied by a barbette in the salient.

s', the area occupied on the terre-plein by the powder magazine.

s'', the area occupied by the traverse and the passage in its rear.

l, the number of yards on the interior crest for a barbette in the salient.

l', the width of the outlet F in yards.

Now as the area allowed for each man is one and a half square yards, we shall obtain the following equation, to represent the relation between the terre-plein, the garrison, &c.

$$(x-9)^2 = 1{,}50\,y + s + s' + s'' ; \quad . \quad . \quad . \quad . \quad (1)$$

this is termed *the equation of the interior space.*

As each man occupies one lineal yard along the interior crest, we obtain also,

$$4x = \frac{y-r}{n} + l + l'; \quad . \quad . \quad . \quad . \quad . \quad . \quad . \quad . \quad (2)$$

which is termed *the equation of defence*, as it expresses the relation between the development of the interior crest, the remainder of the garrison after taking out the reserve, the number of ranks for the defence, and the length of the interior crest required for the cannon in barbette, and for the outlet.

To show the application of the equations (1) and (2), let it be required to find the side of a redoubt, and the number of its garrison, which shall be defended by two ranks on the banquette, after taking out a reserve of one third of the whole number, and have a barbette in each salient.

By a calculation easily made, it can be shown, that each barbette will occupy about 75 square yards of the terre-plein; a powder magazine for four guns, 20 square yards; a traverse of earth, with the passage between it and the foot of the banquette, about 180 square yards; or if of timber about 50 square yards; the portion of the interior crest for each barbette will be nearly 18 lineal yards; and that for the passage about 4 lineal yards.

Taking Eq. (2), and making the substitutions required by the conditions of the Prob. we obtain,

$$4\,x = \frac{y - \frac{1}{3}\,y}{2} + 4 + 72\,;$$

and from it,

$$y = 12\,x - 228.$$

Substituting this value of y in Eq. (1), and placing for s, s', s'', their values, we have

$$(x - 9)^2 = 18\,x - 342 + 300 + 20 + 180\,;$$

or $$(x - 9)^2 = 18\,x + 158\,;$$

from which, by solving the equation,

$$x = 38 \text{ yards}\,;$$

consequently $y = 228$ men;

and $r = 76$ men.

NOTE B.

Manner of equalizing the Excavation of the Ditch, and the Embankment of the Parapet.

Fig. B.

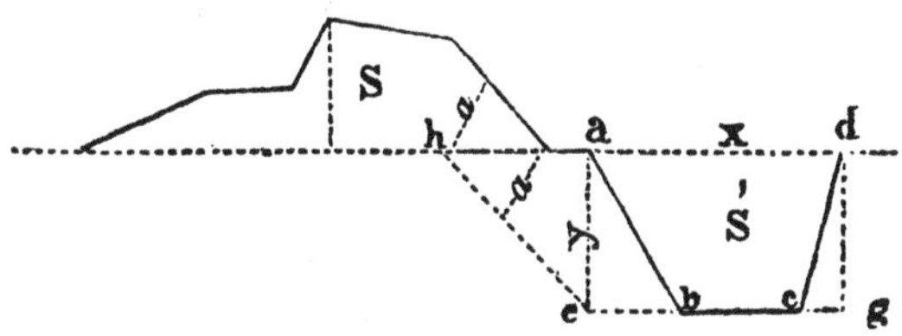

Let the following notation be adopted, Fig. B.

V, Volume of the embankment.

S, Surface of its profile.

l, Line described by the centre of gravity of S in generating the embankment.

V', S', and l', corresponding values of the ditch.

Now, from the centro-baryc principle, we have $V = S\,l$; . . (1), and $V' = S'\,l'$; . . (2).

But as earth when freshly thrown up occupies a greater volume than in its natural state, if this increase be represented by the fraction $\frac{1}{m}$, we shall have to represent this condition

$$V = V' + \frac{1}{m} V'; \quad . \; . \; . \; . \; (3)$$

and eliminating V and V′, by means of these three equations, there results

$$S' = S \frac{l}{l'} \left(\frac{m}{m+1} \right); \quad . \; . \; . \; . \; (4)$$

or the value of the surface of the profile of the ditch in known quantities.

Representing now by

x, the width of the ditch = a d;

y, its depth = a e;

a, the angle of the natural slope of the earth with the ground = a h e.

We obtain

$$\text{a h} = y \; Cot. \; a;$$

and for the bases of the slopes of the scarp and counterscarp,

e b = $\frac{2}{3}$ a h = $\frac{2}{3} y \; Cot. \; a$; and c g = $\frac{1}{2}$ a h = $\frac{1}{2} y \; Cot. \; a$;
we then obtain for the area of the trapezoid a b c d, which is S′, by subtracting the surfaces of the triangles a e b, and c g d, from the rectangle a g,

$$S' = x y - \left(\tfrac{2}{6} y^2 \; Cot. \; a + \tfrac{1}{4} y^2 \; Cot. \; a\right);$$

or, $S' = x\,y - \frac{7}{12}\,y^2\ Cot.\,a\,;$ . . . (5)
an equation from which either x or y may be obtained, by assuming either the width or the depth. But as x has certain limits, from the nature of the Prob., we will solve Eq. (5) with respect to y, and will obtain,

$y = \frac{6}{7}\ Tan.\,a\,(x - \sqrt{x^2 - \frac{7}{3}\,s'\ Cot.\,a},)$ . . (6).

We observe here, in the first place, that the radical expression admits only the minus sign before it, because x and y are decreasing functions of each other; that x^2 cannot be less than $\frac{7}{3}$ S$'$ *Cot. a*, otherwise the expression will become imaginary; and when it is equal to it, the profile of the ditch becomes a triangle. So that the minima limits of x will be,

$x = \sqrt{\frac{7}{3}\,S'\ Cot.\,a}\,;$ and $x = 12$ feet; (Art. 39)
and the maximum value will be found by multiplying the height of the interior crest by the denominator of the fraction which represents the superior slope, and subtracting from this result the horizontal distance between the interior crest, and the crest of the scarp. (Art. 34.)

Any quantity, therefore, which is greater than the minima limits, and less than the maximum, will satisfy the conditions of the Prob., if it gives for a result y between 6 and 12 feet. (Art. 39.)

But should the value of y not fall between these limits, then we shall be obliged to change the form of S; and consequently find a new value of S′; making it greater or smaller, as the case may require.

To show the manner of applying this formula to practice, let it be required to find the dimensions of the ditch of a square redoubt from the following data; the command of the interior crest is 8 feet; the thickness of the parapet 9 feet; the superior slope 1-6th; the angle of the natural slope 45°; and the tread of the banquette 4 feet.

Taking in the first place Eq. (4),

$$S' = S \frac{l}{l} \left(\frac{m}{m+1,}\right)$$

a very simple calculation will show that $\frac{l}{l'}$ will be very nearly represented by the fraction $\frac{3}{4}$, for redoubts between 20 and 40 yards of a side, measured along the interior crest, l' being the middle line of the top of a ditch, which is a mean between the least and greatest limits. The value of $\frac{1}{m}$ is, in ordinary earth, about $\frac{1}{8}$, and varies generally between $\frac{1}{8}$ and $\frac{1}{12}$, so that by substituting these values in the Eq. we have

$$S' = \frac{3}{4}\left(\frac{8}{8+1}\right) S = \frac{2}{3} S;$$

and the value of S, calculated from the data, being 122 square feet, we have

$$S' = 81, \text{ square feet nearly};$$

substituting this value of S′ in Eq. (6), we obtain

$$y = \frac{6}{7}\left(x - \sqrt{x^2 - 189}\right);$$

from which we see that $x^2 = 189$, gives $x = 13$ feet, therefore the quantity assumed for x must be greater than this; if then we assume $x = 15$ feet, and substitute it in this formula, we obtain

$$y = \frac{6}{7}\left(15 - \sqrt{225 - 189}\right);$$

or, $$y = \frac{6}{7}(15 - 6) = 7.7 \text{ feet};$$

which satisfies the required conditions.

It may be added, that with this width of ditch, the line of fire passes 2.6 feet above the counterscarp crest.

In the foregoing calculations, no allowance has been made for the volumes of the barbettes; and the traverse when of earth. To make this allowance Eqs. (3) and (4) will have to undergo a suitable modification, as follows:

The cubic contents of each barbette, in a salient, is nearly 140 yards; and that of the traverse about 170 yards; if then the total cubic contents of all the barbettes, and the traverse, be represented by V'', a given quantity, Eq. (3) will become

$$V + V'' = V' + \frac{1}{m} V' \quad . \quad . \quad . \quad (X);$$

and

$$S\,l + V'' = S'\,l' + \frac{1}{m} S'\,l', \quad . \quad . \quad . \quad (Y);$$

now as V'' is a given quantity, it can be represented by $S''\,l$, in which S'' is found by dividing V'' by l; making this substitution in Eq. (Y), we obtain

$$S' = (S + S'')\,\frac{l}{l'}\left(\frac{m}{m+1}\right) \quad . \quad . \quad . \quad . \quad (Z)$$

for the modified value of S' Eq. (4), to suit this case. The operations for finding x and y, Eq. (6) are the same as already shown.

The calculations here given suppose the plane of site sensibly horizontal, and the interior crest parallel to it; when this is not so, as for example in a defiladed work, the volumes of the parapet and ditch will have to be calculated for each face separately.

The method of doing this, where the site is very irregular, consists in making an unequal number of parallel profiles, on each face, at equal distances from each other, and finding the area of eachp rofile, and then using the following notation and formula,

let $s', s'', s''', s^{iv}, \ldots\ldots s^{n+1}$, represent the areas of the successive profiles, their entire number being $n+1$;

l, the distance between the profiles;

v, the volume between the extreme profiles; then

$$v = \frac{l}{3}\left(s' + 4s'' + 2s''' + 4s^{iv} + 2s^{v} \ldots 4s^{n} + s^{n+1}\right).$$

In other cases, where the site is not very irregular, a mean profile may be found, and the Eq. (Z) be used.

It is usually best, since the dimensions of the ditch cannot be uniform throughout, to adopt a uniform depth, and to vary the width as circumstances may require.

www.ingramcontent.com/pod-product-compliance
Ingram Content Group UK Ltd.
Pitfield, Milton Keynes, MK11 3LW, UK
UKHW041120071026
15587UKWH00009B/16

9 780837 105574